# Lifelines

ELAINE CHIN, MD

# Lifelines

## Unlock the Secrets of Your Telomeres for a Longer, Healthier Life

Figure.1
*Vancouver / Berkeley*

15 16 17 18 19 5 4 3 2 1



Cataloguing data available from Library and Archives Canada

ISBN 978-1-927958-40-7 (hbk.)
ISBN 978-1-927958-39-1 (pbk.)
ISBN 978-1-927958-42-1 (epub)
ISBN 978-1-927958-43-8 (pdf)

Editing by Michael Schellenberg
Copy editing by Michelle MacAleese
Index by Stephen Ullstrom
Design by Jessica Sullivan
Printed and bound in Canada by Friesens
Distributed in the U.S. by Publishers Group West

Figure 1 Publishing Inc.
Vancouver BC Canada
www.figure1pub.com

# Contents

# Acknowledgments

**BOOKS ARE WRITTEN** to inspire us to better understand our universe and to communicate information that will improve our lives.

I wrote *Lifelines* because my patients challenged me to share my knowledge of preventive medicine with more people. I've empowered them to stay healthy and feel younger—every day. So thank you, my patients, for encouraging me to be a better doctor and to share my learning.

Kenneth Whyte, a distinguished editor, has urged me to write a book for many years. I've always told him I don't have the necessary skills. I owe this book to him, for nudging me to commit the time to get my words onto paper. And with some help from those who do write for a living, *Lifelines* has become a personal accomplishment.

My communications specialist, Norah Fountain, taught me how to find my written voice as I began crafting blogs and articles with her support for various magazines and TV health reels. This set the foundation for our partnership to get this book written with "Chinisms," as she calls them, embedded throughout.

Without the help of Kelly Jones, this book would not be as grammatically perfect or publisher-friendly as it is. She set the ground rules to make it consistent and accurate. I owe her a great debt of gratitude for so much patience, which I lack, and for such ruthless attention to detail.

And the magic sprinkles and special touches for saying it with flair come from my friend and communications expert, Bob Ramsay. He made sure I put my best foot forward.

With her immense knowledge of the book publishing world, Heather Reisman offered me crucial insights on how to approach the writing of this book. I've completed *Lifelines* with a sense that it will add real value to the many other health books on shelves thus far.

The illustrations in *Lifelines* first appeared in an iBook sold on iTunes, which was designed and produced by Dale Forder. *Stopping the Clock from the Inside Out* is a wonderful graphical, interactive story about the power of telomeres in extending and improving our quality of life. The opening symbolic piece of artwork was painted by Jean Miller Harding with support from her husband, Kent. All the graphics found here were inspired by this iBook.

All books require a publishing team comprising editors, graphic designers, proofreaders, and a sales and marketing team. The world of publishing is changing, and I'm thrilled to be working with the team at Figure 1 Publishing in Vancouver, British Columbia, led by Chris Labonté and with editor Michael Schellenberg. Together, in partnership, as author and publisher, we have moved in a direction of artistic and content collaboration and also mutual financial risks and rewards.

Behind the scenes is a great team of people who support me at Executive Health Centre, the medical practice I founded over ten years ago. Without the equal commitment of our support team, this would not have been possible.

The trend for wellness is afoot. Forward-thinking organizations recognize that preventive health is the key to economic stability and quality of life around the world. I'm proud to be representing TELUS Corporation as its Chief Wellness Officer. I agree with TELUS's Executive Chairman, Darren Entwistle, who believes that health care transformation is the single most important and biggest social challenge of our lifetime. Joe Natale, President and CEO of TELUS, has said that at the end of our careers we should be able to look back and say we've made an impactful difference in people's lives. What better way than to teach people how to stay healthy? We are certainly on a mission to accomplish this goal.

I'm indebted to the many scientists who have been passionately and relentlessly committed to the scientific research that resulted in the knowledge I'm sharing with you in this book. A special note of appreciation goes to Dr. Lennart Olsson, who has kept me scientifically rigorous.

Thank you to Dr. Elizabeth Blackburn for leading the way with her telomere scientists. I was privileged to meet Dr. Blackburn during her 2013 Toronto visit. We shared some wonderful discussions about the importance of telomere biology and its relevance to preventive medicine. Bob Ramsay summed it up so well in his introduction to her lecture: "The success Dr. Blackburn enjoys could have a profound effect on the lives of . . . well, let me be conservative here and say, pretty much everyone on earth."

Many have said that life is a journey full of peaks and valleys. Certainly, I can attest to that. As you will learn in this book, how you manage this journey will determine the quality of your health. Donald Morrison, respected global business leader, adds an important perspective in this book.

I've reached out many times to my teacher, mentor, and now colleague and friend, Dr. Mary Vachon, for counsel. She has said that when I reach 50 years of age, I will discover wisdom. I certainly believe that this book has been an opportunity for me to share my wisdom with you, my readers.

Our children are always the most truthful. During my moments of frustration in writing, my son, Robert, kept reminding me, "You always say to me, keep trying and never give up." I'm proud to share this book with him as our accomplishment.

*Lifelines* is dedicated to all those who aspire to achieve peak health and are peak performers.

# ONE The Telomere Edge

## Mastering the Aging Cycle

1

# The Foundation of Preventive Care

I'VE SPENT HALF of my twenty-five-year career as a doctor working with patients to predict if they'll get really sick. I use genetics and measure their biomarkers to assess how their bodies are "ticking." I then figure out how they can modify their lifestyles and adjust their diets, supplements, hormones, and medications to reduce their risk of disease.

I practice preventive care. It focuses on preventing you from getting sick rather than treating you after you get sick. It is also called personalized medicine, because when it comes to that ounce of prevention, one size definitely doesn't fit all.

Ten years ago, preventive medicine represented only a tiny part of the health care system, and personalized medicine was pretty much a dream. Today, both are respected, real, and growing in influence. So tomorrow, it's very possible we will be able to significantly slow down the ravages of the aging process.

Not everyone is a believer. Some ask me: "What evidence do you have that preventive care really works? Can you prove that your recommendations actually prevent disease and help people live longer?"

My physician colleagues ask me this, the media asks, and my patients and the public all ask too. My answer? A resounding yes. My supportive proof is the Nobel Prize–winning research into a field that simply didn't exist a generation ago: telomere biology. Over the past twenty-five years, scientists have worked to uncover the incredible predictive ability of telomeres to determine your biological age. Their work was so significant that in 2009 the Nobel Prize in Medicine was awarded jointly to three scientists, Drs. Elizabeth Blackburn, Carol Greider, and Jack Szostak, for discovering how telomeres protect our chromosomes and identifying the importance of the enzyme telomerase in this process.

Together, these three scientists solved a huge mystery in biology. How can our chromosomes be copied completely when our cells divide and how can they be protected from decaying and "dying"?

To answer that question, the scientists had to research the long, threadlike DNA molecules that carry our genes and are packed into forty-six chromosomes. Telomeres are the caps on the ends of the chromosomes, like agelets, the plastic ends on shoelaces.

Drs. Blackburn and Szostak discovered that a unique DNA sequence, called the telomere, protects our chromosomes from degrading. Then Drs. Greider and Blackburn discovered telomerase, which produces telomere DNA. When telomeres become shortened to a critical length, cells "age." On the other hand, if your body is producing enough telomerase, the length of your telomeres won't shorten (sometimes they can actually grow) and automatic cell death, what science calls cellular senescence, is delayed.

The awarding of the Nobel Prize has caused a surge of interest in this field of medicine, and many more research groups are now working feverishly to determine which factors cause our telomeres to shorten before their time and how

**Chromosome**

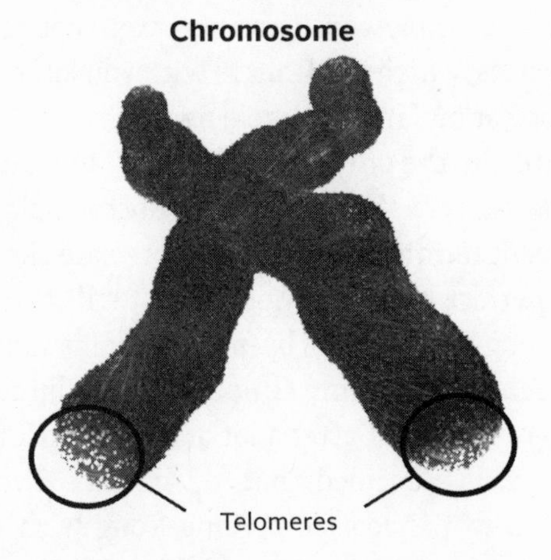


telomeres can be re-lengthened. It will take some years to learn how to re-lengthen our telomeres safely. But today we know much more about why they age prematurely and how we can slow our biological clocks, live longer, and stay healthier.

This is why I'm so committed to speaking out about how important telomeres are in not dying young. Not only can telomeres predict your future health, they can affect how long you'll live and how free of disease you'll be.

That's pretty powerful information. Today, with a simple test, you can learn how long your telomeres are. That one measure alone tells you how much your cells have aged and gives you a measure of your true biological age. You should worry if your test shows that your cells have aged beyond your chronological age, or calendar age, compared to people like you in age and gender. But don't run out to complete your bucket list on the basis of a telomere test. Instead, you should view your results as a chance to look at your life story so far and decide if you need to change a few things.

What's the real promise of telomeres? You can make the right lifestyle choices to ensure that your telomeres stay long and robust. And? You'll have as many years—and as many *good* years—as you possibly can. I can tell you that receiving my own test results, which I will share with you later in the book, made me look at my life in a whole new light.

First of all, I'm the only child of a Tiger Mom and I have the lifestyle and skills to show for it. I'm a doctor, I play the piano, I have a decent tennis swing, and I appreciate the difficulties of being a perfect ballerina. Yet today, I still think I haven't achieved enough—surely the by-product of my upbringing.

Am I resentful about my childhood? Absolutely not. Well, okay, maybe a little. But after a lot of introspection, I've made peace with it. I've decided that, on balance, I've benefited. I immigrated to Canada from Hong Kong with my parents in 1969, when I was five years old. I don't have any regrets that we came to Canada. Far from it. I'm incredibly thankful for what I have and the opportunities my parents have given me.

But when I was a child, I endured a less than ideal family life. My father was a devoted teacher. Because he was laid off a lot during his career, however, I was stressed and worried about our family never having enough money. My mother was a trailblazer. As a young woman, she worked as a senior civil servant in Hong Kong. I wasn't born until she was in her mid-thirties. In Canada, she had to rebuild her career but was able to carry on her public-housing initiatives in Ontario until her retirement. Of course, nobody's perfect—including my mom and dad. Perhaps they could have been better role models when it came to marriage and family life. Working hard is one thing, but moving emotionally away from one another will destroy a relationship. I probably didn't learn enough life skills growing up in that kind of environment, and my marriage unraveled when my son was seven years old.

Being a doctor doesn't make me immune to stress in my personal life or to the strain of parents becoming ill. I watched my elderly father deteriorate over the years from chronic obstructive pulmonary disease (COPD), a lung disease caused by smoking and observed sadly his gradual loss of dignity when he had to rely on help and needed oxygen to move around the house. He eventually died from lung disease. His biggest regret: smoking. My mother suffered from manic depression, now known as bipolar disorder (undiagnosed until I became a doctor), and suffered four bouts of cancer (mouth, breast, colon, and liver) before she finally succumbed to the disease when she was only 72 years old.

Why am I telling you about my parents and childhood?

I hope it will help you understand how genetics and environment act together to determine a person's health destiny.

I believe strongly that by using this book as a guide, you can discover how your past experiences have made an impact on your health and what might be in store for your future. We all start to worry about how to stay healthier as we get older. How about you? What health issues have you faced? What kind of lifestyle do you have? Are you concerned that it has already affected your health?

Until now, all we could do was make an educated guess.

With telomere testing currently available to all of us, you now have an objective, scientific way to find out where you stand—a baseline—and what you need to do to protect that base at all costs.

I've often wondered if my childhood and family life might have caused my telomeres to shorten prematurely. Research shows that chronic stress shortens your telomeres. In a groundbreaking study to be discussed in more detail in Chapter 9, researchers found that mothers who were caring for chronically ill children had shorter telomeres than mothers who cared for generally healthy children. That shortening, in turn, leads to disease starting sooner rather than later. Or not at all. How you cope with the stress can reduce its damaging consequences.

But make no mistake: stress can kill you. Telomere science proves this to be true.

When you were conceived, your biological parents gave you a set of genes—the blueprint—which determine how your body functions. But in most cases, how you live your life and how you perceive it will determine when and if you stay well—or get sick.

A simple principle, isn't it? In other words, a positive lifestyle and a positive attitude can determine pretty much everything.

I can say without a doubt that I live a healthy lifestyle. I never skip meals, and I eat a low-carb, low-red-meat diet, with lots of fish, fruits, and vegetables. I've never smoked (well, I had one drag of a cigarette when I was five years old because it looked so appealing... and my God, never again). I drink little alcohol—one glass socially one to two times per week, max. I sleep well, usually for seven hours (with the exception of when I was on call as an intern in a hospital), and I stay active, but I'm not an exercise freak.

Is that enough for me to stay healthy for my entire lifespan? Is what I do enough to keep my telomeres from shortening prematurely?

Before I got tested and found out my telomere length, I believed (and so did many people close to me) that I had a good chance of having shorter telomeres than the average person because of the chronic pressures in my life. I wondered whether lifestyle and a positive outlook could really compensate for the physical damage caused by emotional stress.

As you read further into this book, I encourage you to complete your own self-audit. See how you measure up. Perhaps you've burned the candle at both ends at some point in your life and worry about that coming back to hurt you. Don't fret. Instead, let's figure out what you can do now to make up for your past.

I'm always as intrigued as my patients are to learn about their telomere results. Certainly, when patients get a better result than they expected, I see on their faces how relieved they are that they've "gotten away" with it. Conversely, for those who get unexpectedly poor results, we have to discover what led to such outcomes. Given either result, the test is always a motivation to improve one's lifestyle behavior, especially to stop smoking, drink less, and exercise to achieve a healthier weight. No one likes to lose their telomere length prematurely. The common driving force to change behavior is to stop tempting fate.

Deep inside our psyches, many of us would like to live forever—so long as we're of sound mind and free of disability and disease. Who wouldn't want to achieve this health goal?

If you want to know how to start on the path of living longer now and can make the leap of faith in the new science of telomeres, this book is for you. After going over the reasons why I have devoted my life to following this science, I will give you the keys to the Peak Health Program. I'm going to show you how to assess your baseline measurements so that you can determine what you need to include in your day-to-day routine to live in a state of excellent health. I have patients right now who have tested their telomeres and are already living better and healthier. Not only are they "feeling" the rewards of even tiny lifestyle changes, they look forward to "seeing" the proof for themselves the next time they test their telomeres. It's why I've written this book—so you too can follow the program and know what it takes to be healthy and stay that way, from your lifespan right down to the length of your telomeres.

# 2

# Welcome to the New Science of Aging

*Telomere maintenance is tied to the reasons why most people die... understanding [telomerase] may eventually be the basis for therapies to combat cancer, heart disease, and diabetes—perhaps even halt the ravages of age.*[1] DR. ELIZABETH BLACKBURN, Nobel prize–winning scientist

## Aging through History

Our desire to stay young and vibrant, live longer, and continue to perform at our peak has been an ideal since we first walked the earth. The Greeks created the foundations of our modern Olympics, challenging a man's strength and speed and celebrating exemplary physical performance. Cleopatra of Egypt perfected the art of makeup. Indeed, ancient Egyptians' concerns with beauty and body care transcended gender lines. Women and men both used cosmetics and body oils. And for thousands of years, Chinese herbalists have created tinctures for vitality, mental acuity, and sexual prowess.

But I'll let you in on a secret: there has always been a Fountain of Youth—we just didn't know where to look for it. Now we do.

The twenty-first century has seen the birth of the professional athlete, one who gets paid millions of dollars to win. Many have turned to science to help enhance their performance, with coaches, psychologists, trainers, and medical professionals. Athletes watch what they eat, take supplements, and—yes—inject hormones and drugs too. Whether fair or legal, these techniques work to enhance the body's repair and growth mechanisms that help athletes perform better. In the words of science, we are essentially manipulating our physical DNA expression.

We've become more scientific in how we ingest our tinctures—what we now call vitamins and supplements. Their use has skyrocketed in the last few decades. In fact, the United States Centers for Disease Control and Prevention reported that more than half of American adults used supplements regularly.

Many people who take vitamins and supplements do so because they believe that consuming them will make them look and feel youthful and energetic, and will help them enjoy longer, healthier lives.

Hormone-replacement (HRT) therapy is another way that some men and women enhance their health. Women take estrogen and progesterone in menopause, and middle-aged men consume dehydroepiandrosterone (DHEA) and testosterone to treat andropause symptoms. Enhancing diminishing sexual performance through prescription drug use is now socially acceptable too, with sales of Viagra in 2010 topping nearly $2 billion, not to mention sales of Cialis and Levitra.

We can use topical treatments to enhance our health as well. Sunscreen protects our skin from premature aging. We have all types of skin care for women and men. And some people elect to move on to more invasive techniques, such as injectables (Botox and fillers), chemical skin peels, laser skin resurfacing, and cosmetic surgery. As we're living longer, we're working harder to look—and feel—younger. We also worry about spending our last years in a hospital bed.

> The lag between healthspan and lifespan is serious. None of us look forward to chronic illness. But as centenarian George Burns put it: "Old is ten years older than I am."[2]
>
> TOM EGAN, Chair, Sefton End Ageism Group

## Lifespan

The fact is, we have added decades to our lifespan in the last century. In *The Improving State of the World: Why We're Living Longer, Healthier, More Comfortable Lives on a Cleaner Planet*, author Indur Goklany notes that life expectancy before 1820 was just 20 to 30 years. Since 1900, he points out, our lifespan has risen from 47 to 78 years. Thanks to environmental and medical advances since that time, we are, in essence, living beyond our evolutionary function to procreate our species. I say this because life expectancy analysis from the World Bank in 2013 shows the life expectancy of a child born in Japan is now 83 years. And if you're a baby girl born in 2013 in Tokyo, the chances of you living to be 100 are now one in two.

In the United States, life expectancy is pegged at 78 years. In Canada, it's 81. To see the life expectancy rate for different countries, visit the World Bank website at http://data.worldbank.org/indicator/SP.DYN.LE00.IN.

DEFINITIONS

**Average Lifespan** and **Life Expectancy** The terms *average lifespan* and *life expectancy* refer to the average number of years that a person has left to live at any given age. This contrasts with *maximum lifespan*, which is the documented upper limit of years of life for the oldest people alive anywhere. The maximum human lifespan is between 120 and 130 years. Yes, a few people have lived to be 130. But only very, very few.

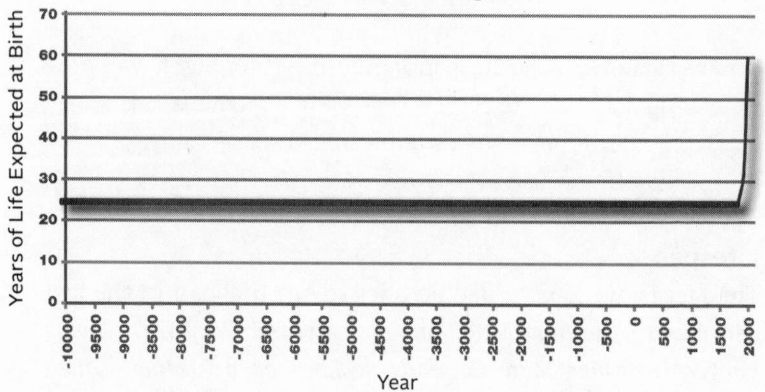


Still, despite some astounding scientific advances, our maximum lifespan hasn't changed all that much in the last sixty years. It seems to have peaked. In fact, the expected lifespan of North Americans has only risen 15 per cent, and the current rate of increase has been steadily slowing since the 1950s.

Why the plateau in how long we live? I believe it's mainly due to how we've focused on treating disease over the past sixty years. It's myopic and limited.

Why? All the trillions of dollars in research money invested in disease treatment no longer have the significant gains of a century ago. So we need to focus our money, skills, and energy more on wellness—and less on illness. We need to find the root cause of age-related diseases such as heart disease, cancer, arthritis, and some forms of dementia.

Post–World War II, we had a technology boom in medicine. Today, we can literally see the development of cardiovascular disease using ultrasound, CT, and MRI scans. As well, we have a clear understanding of what causes and how to manage type 2 diabetes. Billions of dollars are being spent on monitoring blood sugar (such as with home glucometers) and cardiac rhythms (with pacemakers and auto defibrillators). We

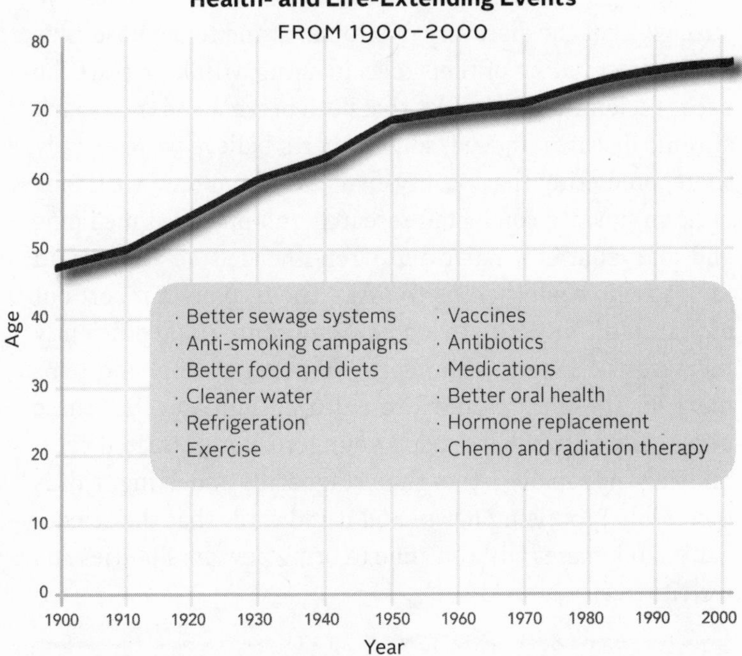


diagnose cancer with reasonable certainty. Arthritis is now managed with many new biologics (customized drug therapies). Certainly we are a society managing chronic disease with prescription medications and chemotherapies. This is big business. Only a handful of companies stand to gain from these technological advances. Certainly the incremental gain in quality life years is much less in the second half of the twentieth century compared to the first half.

For all of us living into the twenty-first century, larger gains in quality life years will not be found in more drugs and procedures. We need to discover and act on the causes of premature telomere shortening which leads to disease. Some of those actionable solutions are in this book. It doesn't cost you much, just the firm decision to carry through with a change in behavior and life attitude.

The good news is, we're getting oh-so-close. I believe strongly that the discovery of telomeres and telomerase, plus our understanding of their roles in aging, will kick-start the pace of scientific discoveries that both prevent and treat many chronic diseases. Indeed, some experts believe we're already there, predicting that twenty-five years from now, we'll look back on how we conducted research and practiced medicine and marvel at how naive and wrong-headed we all were. In fact, as this book is going to press, there comes a report out of Stanford University that a new procedure could efficiently increase the length of human telomeres by applying telomere lengthening proteins to cultured human cells. These petri-dish cells behave much younger than untreated cells, multiplying rapidly rather than stagnating and dying. This is incredibly promising news—and it indicates that the science behind telomeres will continue to bring new possibilities and rewrite treatments.[3]

SCIENCE STAT

**What's in a Telomere?** A telomere is a section of nucleotide sequences at the end of each of our chromosomes. (Nucleotides are organic molecules that form the basic structural unit of nucleic acids, such as DNA.) Each section is made up of more than three thousand copies of the sequence TTAGGG. The repeated sequences protect the ends of chromosomes from unraveling, like aglets on a shoelace. Without them, our chromosomes fuse together and degrade prematurely causing our cells to die, which leads to us aging before our time.

### The Future of Funded Research

Money will follow this area of research because the baby boomers not only want their quality of life to remain stable, they'll insist on it.

We know that wishing won't make it so. But the combination of money and will and technology could. It's very possible that technology, which has led to so many miracle cures in the past decade alone, will find a way to slow the aging of our cells. If this happens, we'll be able to cure the biggest killers in the world today (heart disease, diabetes, and cancer) and prevent disabling conditions like arthritis and dementia from developing at all.

Again, this *could* happen.

Now for the money and the will. As a group, the boomers have amassed a great deal of wealth and plan to stay alive for a long time to enjoy it. They are willing to invest their cash into this area of medicine.

Proof that big money has entered the fray of curing disease is Google's new venture called Calico. It will "focus on health and well-being, in particular, the challenge of aging and associated diseases." In the words of Google's CEO, Larry Page: "Illness and aging affect all our families. With some longer-term, moon-shot thinking around health care and biotechnology, I believe we can improve millions of lives."[4]

In September 2014, Calico announced that it was partnering with Chicago-based pharmaceutical giant AbbVie. It is a $1.5 billion commitment to develop life-enhancing therapies for people with age-related diseases.

## The Science of Telomeres

This book is about your ability to stop the ravages of aging. It reveals how all of us can use the Nobel Prize–winning science of telomeres to become more youthful, energetic, and healthy. Telomeres stop your DNA from unraveling. Medicine now knows how to slow that process—and even correct it—in order to turn back the biological clock and prevent disease. For the first time in the history of medical science, we can tell "cellular" time. Like the rings of a tree trunk, our cells are telling us their age through the length of their telomeres.

In fact, our telomeres function as our biological clock. We know mechanical clocks can be reset, so why not our biological clocks? It's not that simple, of course. But this is the essence of anti-aging medicine—slowing down this ultimate regulator and finding ways to potentially turn back the clock from inside our bodies by modifying our lifestyle and manipulating our biochemistry, which slows down the shortening and/or lengthening of our telomeres.

The study of telomeres is a science in its infancy. Research papers about telomeres have been published only in the past twenty years. Yet in this short time, telomere research has given us an immense new understanding of how we age and why we develop chronic diseases.

> Giving a Nobel Prize [for telomeres] has made this field of anti-aging more acceptable. It has now become a real focus in peer-review science.
>
> WILLIAM ANDREWS, PhD, Founder, Sierra Sciences

True, scientists are studying telomeres to quench our universal desire for an elixir of youth. But more important, they want to stop the greatest killers the modern world has ever seen—cardiovascular disease, cancer, Alzheimer's disease—and to halt debilitating illnesses such as osteoporosis, arthritis, and liver cirrhosis.

Here's an even greater possibility created by our new understanding of telomeres. There is strong evidence that telomere shortening may be the root cause of all of these conditions. What's more, scientists are making an ever-stronger case that there's not only a direct relationship between the length of your telomeres and the length of your life, there's a causal relationship. In other words, short telomeres may actually *cause* you to live a shorter life by making you vulnerable to disease, and long telomeres may *cause* you to live a longer life by protecting you from disease. When it comes to telomeres, I say "length matters," and the longer the better. So the race is on to discover how to lengthen telomeres. And one thing is certain at this early stage: the enzyme telomerase is deeply implicated in this process.

What we learn from telomeres and how to keep them going strong gives us a road map for preventing premature illness caused by diseases that can lead to all of us dying before our time. Today, for the first time, we can actually monitor, manage, slow down, and perhaps even reverse the effects of the aging process.

## How Old Are You, Really?

We are eternally curious about how old other people are.

Although we rarely ask people their age directly, when we do find out, we're often surprised by their answer: we might think they appear older or younger than their chronological age. The average person wonders, "How old are you?"; telomere scientists might ask, "What is your biological age?"

Physicians always ask patients about their age because a patient's age tells them many things about their risk for certain types of medical conditions. It's obvious that some diseases occur more commonly in childhood due to infection (such as croup and scarlet fever), and other conditions develop in later life. But what causes those conditions has not always been clear, especially with diseases such as cancer and heart disease—until now.

With the discovery of telomeres, our chronological age, or calendar age, now seems like a very blunt measure of our biological risks. What we need to know in medicine today is how well we are aging in terms of our telomere length. This, in turn, gives us a much more accurate marker of our health and life expectancy—in other words, our biological age.

### BLUE ZONES

Many illnesses develop as we age, and quite a number of these can be prevented. Using hard data, researchers like Dan Buettner, an American explorer, educator, and author, have discovered there are specific regions of the world where people are living longer and healthier.[5] In 2008, Buettner released a book on his findings, *The Blue Zones: Lessons for Living Longer from the People Who've Lived the Longest*, through National Geographic Books. Buettner calls these regions Blue Zones, and happy, healthier 100-year-olds are the norm in such top lifespan spots as:

1. Ikaria, Greece
2. Nicoya, Costa Rica
3. Okinawa, Japan
4. Sardinia, Italy
5. Loma Linda, California, USA

Many people who live in these places share some similarities in lifestyle. In upcoming chapters, I explore the role that active living, diet, and outlook play in shaping telomeres—and, in turn, shedding light on the secrets to longevity.

FURTHER INFORMATION

See the NBC news story "The Island Where People Forget to Die." bluezones.com/2012/12/nbc-news-the-island-where-people-forget-to-die.

---

## Living Your Full Lifespan Potential

Many scientists believe that genetics only contributes 50 per cent to our longevity and that environmental factors determine the remaining 50 per cent. This means we should be able to modify our lifestyle to improve our longevity.

**Telomere Burnout**

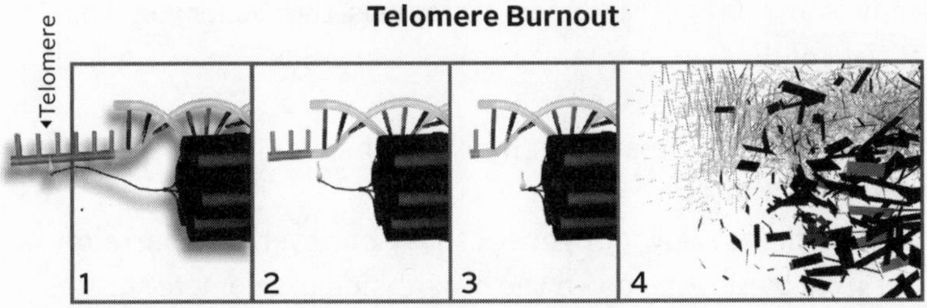

Telomere

1 2 3 4

Imagine the telomere as a wick burning on a stick of dynamite. When the wick burns out, the dynamite explodes. When our telomeres reach a critical short length, disease risk begins.

Your biological parents gave you a set of genes, and these genes will play a major role in determining your lifespan. Your genes are housed in forty-six chromosomes, and the tips of them—your telomeres—are also inherited. Genetics determines 80 per cent of the variation in the length of your telomeres when you're born, but the role played by your parents in the variability of their length diminishes very quickly with time. In fact, that process starts right after you're conceived.

Scientists study identical twins to determine the roles that "nature" and "nurture" play in our development. You won't be surprised to learn that, although identical twins are born with the same number of telomere base pairs, researchers now know that environmental factors play the biggest role in telomere length after birth.[6, 7] In other words, telomere length in twins is the same at birth, but they shorten at different rates depending on the experiences and lifestyles of the individuals.

Initially, researchers believed that short telomeres were the *result* of aging—now we know telomere shortening is the *cause* of aging. So it follows that what you do with your life ultimately determines if you live out your genetic potential. It also can determine which diseases you'll contract as you age: the more you degrade your telomeres with poor lifestyle choices and physical and mental traumas, the sooner you'll be at risk for illness.

### THE RELATIONSHIP BETWEEN TELOMERES AND CHRONIC DISEASE

We can change how our genetics play out over the course of our lives by slowing down the progression of telomere shortening and by re-lengthening our telomeres. Lifestyle changes play a role here, and I cover all you'll need to know about how to make those changes. But first, let's ponder what could happen if you ignore the science of telomeres. Just consider the relationship between telomeres and these chronic diseases.

*Alzheimer's Disease and Dementia* Much shorter telomeres have been found among people who developed an aggressive decrease in brain function in both of these conditions.[8, 9]

*Atherosclerosis (Vascular Disease)* In atherosclerosis, which can lead to heart attacks and strokes, scientists have discovered that the blood vessel cells and smooth muscle cells next to areas of plaque formation have shorter telomeres than healthier counterparts in other areas of the body.[10] Another study found that shortened telomere pairs corresponded to a 300 per cent increase in the risk of myocardial infarction and stroke.[11, 12] There's also a strong correlation between short telomeres and vascular dementia.[13]

*Cancer* Many cancers involve chromosome rearrangement and mutations, and telomere shortening boosts the incidence of both.[14] This is why most cancers develop in older age.

*Infections* Your immune system is highly sensitive to the shortening of telomeres. Its effectiveness depends on the ability of your white blood cells to proliferate during times of immunological stress. It turns out that if you have shorter white blood cell telomeres, you have an 800 per cent higher mortality rate from infectious diseases.[15]

*Liver Cirrhosis* Hepatic (liver) cells have shorter telomeres. When the liver can't repair or reproduce its cells, it grows stiff and malfunctions.[16]

*Osteoarthritis* In osteoarthritis and degenerative disc disease of the spine, cartilage cells have shorter telomeres. This results in a decreased regrowth of cartilage in joints and this results in less cartilage as we age, and more grinding of bone on bone, causing pain and inflammation in our joints.[17,18]

*Osteoporosis (Bone-Thinning Disease)* Bone cells called osteoclasts remodel our bones, and other bone cells called osteoblasts rebuild our bones. In osteoporosis, the osteoclasts keep remodeling, but osteoblastic cells fail to rebuild. Science has discovered a link between the incidence of osteoporosis and shorter telomeres in osteoblast cells.[19,20]

*Wrinkles* Although not life-threatening, wrinkles are a condition we all experience as we age. But why are some people more wrinkled than others? Short telomeres have been linked to the malfunctioning of the basal cell layer, which causes a thinning of skin and age spots.[21]

### PREVENTIVE CARE TRUMPS DISEASE MANAGEMENT

I've said it many times before: our health care system needs to change from managing disease to preventing disease. And knowing how telomeres fit into the preventive health mix is definitely an anti-aging breakthrough. As of now, it's possible to slow our telomere shortening by changing our lifestyle to reduce inflammation, glycation, and oxidative stress, and to improve the balance of our hormones. More on these preventive steps later.

But these steps aren't as powerful as actually lengthening our telomeres as we age. Just imagine the possibilities for our health if we could do that. Well, that day is not far off. Scientists around the globe are working to discover how to activate the enzyme telomerase. The secret to re-lengthening our telomeres may lie in our ability to alter the activity of this enzyme.

Telomerase is an enzyme that protects our DNA from unraveling and mutating. If we could get it to work on our behalf, we could live much longer and be free of many of the diseases that kill us now.

The good news is, it turns out that many animals already have this ability to activate telomerase, such as lobsters, clams, turtles, whales, bats, and starfish.[22] Most of these species don't die of old age, but of trauma or infection.

There's also the Leach's storm petrel, a bird so small it can fit into a human hand. One study showed that this bird lives longer than other birds—more than 30 years.[23] It's also the only known animal whose telomeres grow longer with age. My sense is that these are not freak examples from the animal world; instead, they are valuable entry points into discovering more about the nature and potential of telomerase.

*Starfish Do It. Birds Do It. Why Can't We Do It?* Until we can discover and harness the power of telomerase activators, we must rely on slowing down the rate of our telomere length shortening. In other words, we essentially have to work with what we've been given. I have talked about how telomere length decreases as you age. Certainly we know this now: the faster you burn out your telomeres, the sooner you could become sick. The next challenge—and, in fact, the new era—for anti-aging medicine is all about optimizing our ability to slow down the processes of aging and, in turn, the potential for disease. This means we need to focus on these three big preventive steps, which I also explore in subsequent chapters.

1. Reducing inflammation: Inflammation is a protective response by the body to rid itself of the harm caused by pathogens (such as bacteria and viruses) or irritants. In this process of repair and healing, tissues can be damaged, which leads to damaged organs and disease.
2. Preventing oxidative stress: Oxidation occurs when an oxygen molecule binds to a substance such as organ tissue and causes a change in its chemical structure, thereby creating an unstable, destructive molecule known as a free radical. Long-term damage by these molecules can lead to DNA damage, loss of cellular energy, and—ultimately—cell death.
3. Avoiding glycation: Glycation weakens your body's biomolecules. It binds sugar molecules to lipid or protein molecules inside your body's tissues. This causes your organs, such as your kidneys and even your eyes, to malfunction. Glycation is implicated in many age-related chronic diseases, such as cardiovascular diseases, Alzheimer's disease, and cancer.

By measuring and managing variables such as inflammation, oxidation, and glycation, you can slow down the destruction of your telomeres. Of course, each of us is different and will respond to certain techniques better than others. Thankfully, it's possible to measure the rate of shortening using a simple test. So now you can determine if any of the interventions and treatments I recommend in this book are working for you. Welcome to personalized anti-aging medicine.

### PUTTING NEW SCIENTIFIC DISCOVERIES TO WORK FOR YOU

In the second part of the book, I look at ways to slow down aging by delaying the shortening of your telomeres in six related categories:

1. Nutrition
2. Supplements
3. Hormones

4. Medications
5. Lifestyle
6. Stress management

It sounds like a lot to take in at once, but you'll be amazed at how small changes can lead to big improvements in your health—and to the slowing down of your biological clock.

> We need to change our approach and address the root cause of age-related disease, which will reside within the mechanisms of aging. Not by better treatments for disease but by maintaining health and preventing diseases from occurring in the first place.
>
> CALVIN HARLEY, PhD, in Drs. Elissa Epel and Calvin Harley's TedTalk, "Telomeres and Aging" at TedMed 2011[24]

# 3

# What Are Telomeres?

## Telomeres, Telomerase, and Aging

Apart from death and taxes, there's not a lot that's certain in life.

What else can we count on? The moon will rise, the sun will set, and each person on earth is a unique being. Even if you've never taken a biology course, you know that every one of us is different because of DNA. Countless forensic cop shows have made DNA a common theme, even though the term *deoxyribonucleic acid* itself doesn't roll off the tongue as easily as *DNA*.

What's not as well known—and I'm aiming to help change that—is that, just as aglets determine the stability of shoelaces, telomeres determine the stability of our DNA strands and therefore determine how long we'll live. We want to shout "telomeres" from the rooftops and make them just as well known as DNA.

Earlier, I compared telomeres to aglets. There's an important difference to understand in this simple comparison. When a shoelace breaks down, we just buy another. But when a telomere can no longer do its job, the chromosomal material it's protecting degenerates—and so does a part of you, aging your body and bringing you closer to chronic disease and possibly death. Telomeres are essentially your true biological clock, a major determinant of how long you will live and be healthy. People have often asked me: "So that's it? If I have too many short telomeres, I'm doomed?"

The answer is no.

Let's use a different analogy to help explain the real risk of telomere shortening. How quickly we wear away the treads on our car tires depends on a number of factors: how far we drive, the pressure in the tires, and how often and hard we slam on our brakes. (In the coming chapters, I'll describe what wears and tears your telomeres.) Once our tires become bald, we can still drive on them, but they are less stable and less reliable in wet weather. They can increase our risk of getting in an accident if they have a poor grip on the road. Ditto with short telomeres: they make our chromosomes less stable and put us at higher risk for developing abnormal cells and disease.

DEFINITION

**Telomerase** Telomerase is an enzyme that carries its own ribonucleic acid (RNA) template for DNA synthesis. In the language of science, it is what is known as a reverse transcriptase—that is to say, it is an enzyme that catalyzes the formation of RNA from a DNA template during cell replication, or transcription.

In theory, it may be possible to induce the human telomerase gene to produce telomerase (recall this enzyme can repair telomeres). Imagine the possibilities of that. Having the ability to prevent cells from aging would forever change the landscape of medicine—and, really, life as we know it. This sounds dramatic, and it is.

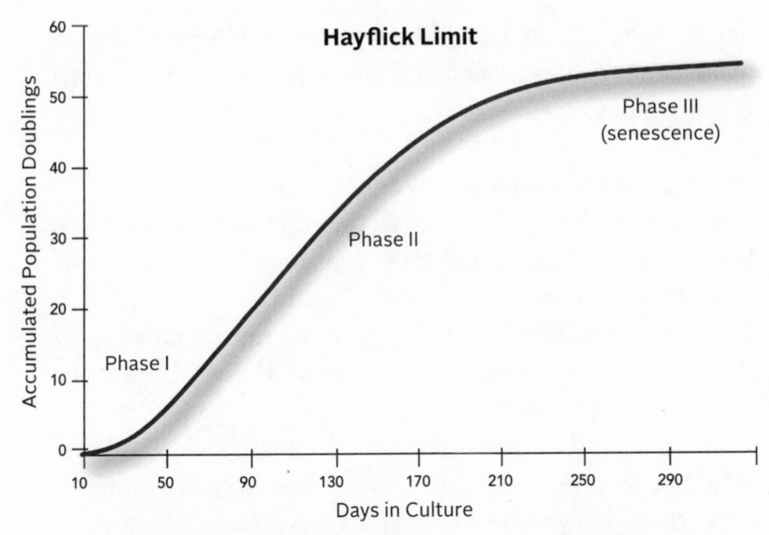


This graph shows cellular division over time. For this skin cell growing in a petri dish, the maximum number of divisions is approximately 50. The maximum age of this cell is approximately 290 days.

As it stands now, each human cell divides a certain number of times before entering a nondividing state called replicative senescence. The number of times a cell can divide is called a cell's Hayflick limit. Different kinds of cells have different Hayflick limits (skin cells in a petri dish, for example, can divide approximately fifty times before they stop dividing). When enough of an organ's cells have died, the organ begins to function inefficiently. This leads to organ failure, which eventually can lead to the organism dying. The discovery of telomeres supports the Hayflick limit theory. We now know that this mechanism of telomere shortening causes cellular senescence, or cell death. However, cancer cells divide indefinitely, defying the Hayflick limit. Today we know this is caused by the continuous activation of telomerase within unhealthy cells.

So it seems if we could kick-start our body's production of telomerase, we could extend the Hayflick limits of all our healthy cells and reap the benefits of robust telomeres indefinitely. That would change the nature of medicine completely—but it's a very big "if."

### What Is Your True Biological Age?

Note that the heading of this section includes the word *biological* rather than *chronological* age. Just because we now know that telomeres exist doesn't mean you have to give up birthday celebrations. To understand the role telomeres play in your biological aging, we need to go all the way back to the moment of your conception.

At that very instant, your telomeres were plentiful and long. In telomeres, length *does* matter. But as soon as conception occurs and you start to develop as a fetus, some of those telomeres begin to shorten and over time are then lost—the telomere version of decaying. Lots of cell divisions occur during this phase of your growth. As you grow older and

**Telomere Shortening**

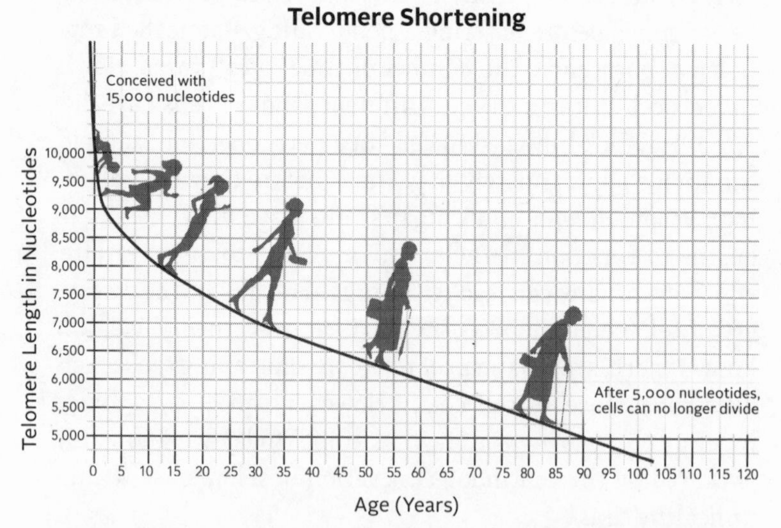


From the moment we are born, we begin to "burn up" our telomeres. At birth, we start with tens of thousands of telomeres (measured in nucleotide units), and by old age, we have fewer than five thousand.

**Your Telomeres**

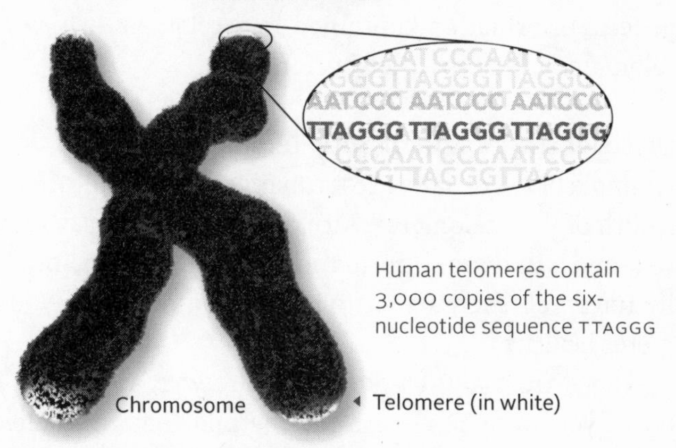


become an adult, the pace of growth slows, and by age 20, your body will likely enter into a repair phase of cellular life, fixing the damaged cells that have been affected by wear and tear, both natural (like endurance sports training) and accidental (like recovery from major surgery from an injury). Although we know that telomerase activates throughout your entire life, that activation is pretty small as you grow older, and the net effect is that your telomeres shorten continuously.

We may not like to think about it, but we know that just as we age every day, we're also dying. But maybe we didn't realize we begin dying the moment we're conceived.

Knowing that our telomeres are already shrinking, we also know we have some work to do if we're going to make up for what we've already lost. We may not be able to reactivate a telomerase gene in humans yet, but activating the telomerase enzyme can be achieved now—even though it isn't ready for prime time in people yet.

### What Does a Telomere Look Like?

The length of each telomere on our individual chromosomes is different. What's more, it's the shortest telomere that determines the lifespan of a cell. Unlike DNA, there's no genetic information contained in a telomere, which is why telomeres are often called buffer DNA.

### Discover the Length of Your Telomeres

A simple blood or saliva test can give a good indication of the health of your telomeres. Although the test is easy for you to take and will reveal information you and your doctor can easily interpret, there's a lot to do afterward to keep your telomeres healthy.

There are many ways to measure your telomeres. Obviously, you want to choose the method that will deliver the most relevant insights for you. There are pros and cons to each testing method, beginning with ease of access, collection of samples, and costs. Most methods reveal average telomere length, and some reveal your percentage of short telomeres.

For many people, calculating the percentage of short telomeres is deemed to be the gold standard because researchers now understand that critically short telomeres in individual cells is the determining factor in how our cells age and how we develop age-related diseases.[1] There are a number of techniques to test telomere length. The qPCR (quantitative polymerase chain reaction) method of testing average telomere length is the most cost-effective (in the range of $250 to $500). While the FISH (fluorescence in situ hybridization) method can measure both average and percentage, it is much more costly (in the range of $1,000 to $1,500).

I generally tell people to opt for the less expensive test; then, if they are concerned by the results because they fall below the average, they can do additional testing to assess the percentage of short telomeres.

HOW TO READ YOUR AVERAGE TELOMERE LENGTH TEST

Let's review someone's telomere test.

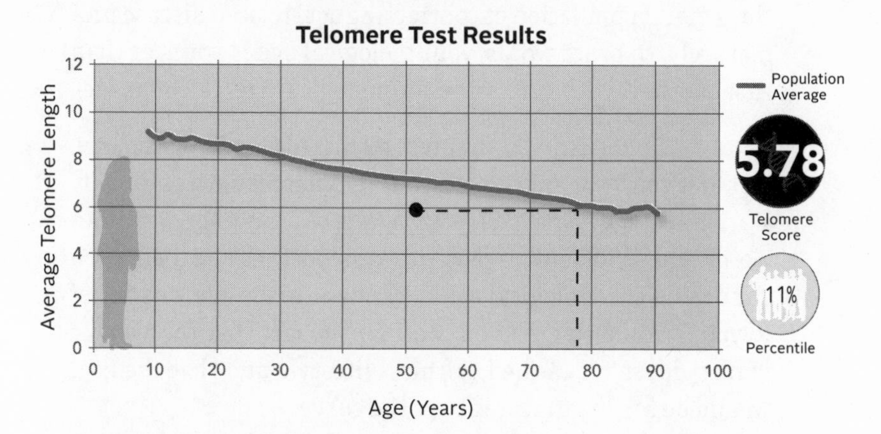


The gray graphed line that runs from age 10 toward age 100 represents the average telomere length for a sample group of approximately seven thousand people at certain ages. This person is 52 years old. Her result, noted as a dot, shows a telomere length score of 5.78. Her result sits below the average curve. For her age group, she is in the 11th percentile. This means that 89 per cent of people in her age group have longer telomeres than she does, and only 11 per cent have shorter telomeres. Therefore, this patient's telomeres are biologically "older" than those of the average person with the same chronological age and sex.

In fact, this 52-year-old woman has cells similar to those of a 75- to 80-year-old. Therefore, based on her telomere results, her biological age is in the seventies.

For your own test, there are three possible results:

1. *You're Above the Line.* If your test results place you above the line, it means you haven't been burning your telomere length as quickly as the average person of your age group and

sex. The higher your telomere score, the biologically "younger" your cells are. This might indicate that you are at less risk than the general population of contracting age-related disease prematurely. In other words, your biological age is younger than that of others with your same chronological age (calendar age).

*2. You're On the Line.* If your test results place you on the line, it shows you've been burning your telomere length as quickly as the average person of your age group and sex. In other words, your biological age is the same as that of others with your same chronological age. Although this might be a relief to you, I would suggest you consider the fact that the average person doesn't lead the healthiest lifestyle and that the goal should be for you to score above the curve.

*3. You're Below the Line.* A below-the-line score suggests your life so far has caused your telomere length to shorten more than the average population's. Your biological age is older than that of others with your same chronological age. Scientific findings suggest that if you are in the 25th percentile or lower, you may develop age-related disease earlier than others of your same age. This is a wake-up call that you need to get moving on improving your health—and quickly! Clearly, the patient in the above chart is below the line at age 52, so she needs to make some real lifestyle changes.

### HOW DO I SEE MY BIOLOGICAL AGE?

Align a ruler horizontally at your test score and draw a line along the X-axis (the age axis). The line you draw will intersect with the gray graphed line at some point. You can find your biological age by following the intersection point down to the age axis.

## Connecting Your Telomere Length Test to Your Health Action Plan

With this new view of your health from a genetic level, you can now create an effective and sustainable health action plan containing specific activities that will improve your telomere health. But to help you achieve even better health, we need to look at the bigger picture.

If you scored above average, this is good news, but you must continue to maintain a healthy lifestyle. Continued monitoring and habit improvements will keep you ahead of the pack.

If you scored average, we should look at your lifestyle, habits, and disease risk biomarkers so you can improve your score at the next telomere test—usually six months after the initial testing.

If you scored below average, it's critical to understand why you aren't maintaining a healthy telomere length. A close examination of your overall health and lifestyle may reveal areas for improvement and you can develop your own health action plan to start improving your prospects and your life.

## What Shortens Telomeres?

Which behaviors shorten lives? We already know a lot of them. Smoking cigarettes. Overeating. In fact, overdoing anything, like drinking or even over-exercising, can negatively affect how long a person will live. Eating the wrong things—such as processed sugar, the world's most "imperfect" food—can lead to glycation. Other foods we ingest and activities we take part in can lead to oxidation or inflammation. Both of these processes can lead to organ damage, which I discuss in more detail in the next chapter. Quite simply, if you don't smoke or overindulge, if you eat right and exercise in moderation, your telomeres will reflect that positive behavior.

DEFINITION

**Progeria** Progeria is a rare genetic disorder that causes babies to be born with extremely short telomeres. People with this condition age rapidly and usually live only until their mid-teens or twenties.

But if you behave "badly," doing all the things we know can lead to getting sick, you're shortening your telomeres. That's bad news. Short telomeres are linked to many chronic diseases, which I discuss in Chapter 1. The good news is that although we're all born with different telomere lengths, our cells will divide at different rates: you can slow down the shortening of your telomeres by exercising more regularly, eating a balanced diet, and taking the right types of supplements, and perhaps even undergoing hormone-replacement therapy.

### What Should You Do If Your Telomere Test Comes Up... Short?

Read the Peak Health Program, detailed in the following chapters on lifestyle, diet, and mental health, and start carving out your own health action plan to make changes to your day-to-day routine. Your goal is to slow down your telomere shortening (in order to set your biological clock back), or hopefully lengthen your telomeres.

What does that? Positive behaviors can stop premature shrinking, but it's the enzyme telomerase that is key to maintaining your telomere length. Every cell can make telomerase, and telomerase can replenish the cell if there's enough of it and if it's "awake" on the job. In some ways, it may be possible to activate your telomerase naturally, either via lifestyle changes or by somehow signaling the cell that it needs to take action to make telomerase.

### How Do We Activate Telomerase?

This is the billion-dollar question that research labs around the world are trying to answer. If they can develop a "pill" that either maintains or even lengthens our telomeres, they can "cure" a number of illnesses, such as cancer. Some forms of cancer are, in essence, a group of cells with no finite life. They are ageless because of an over-activation of telomerase. Certainly, we'd like to understand how to activate telomerase in just the right amount and at just the right time to save our telomeres from shortening due to what doctors call a sudden life insult. By *insult*, I mean any number of medical issues, such as a virus, radiation, a chemical, or a toxin that attacks our cells and, in turn, our DNA.

In 1998, scientists proved it's possible to extend a human cell's natural Hayflick limit.[2] The researchers prompted the cells to produce telomerase in a petri dish, which resulted in longer telomeres and cells that appeared young and healthy. The cells continued to divide normally and exceeded their Hayflick limit by at least twenty doublings. The big surprise was that the cells didn't become cancerous—a big concern of scientists in this field of study.

As is standard in such studies, a control group of similar cells was evaluated under identical conditions and was found to age normally, with telomeres shortening and with expected Hayflick limits.

SCIENCE STAT

**Endless Possibilities** In the January 1998 issue of the *Journal of the American Medical Association*, Professor Michael Fossel, MD, PhD, summarized the real possibilities of telomerase.[3]

> Recent research has shown that inserting a gene for the protein component of telomerase into senescent human cells reextends their telomeres to lengths typical of young

cells, and the cells then display all the other identifiable characteristics of young, healthy cells. This advance not only suggests that telomeres are the central timing mechanism for cellular aging, but also demonstrates that such a mechanism can be reset, extending the replicative life span of such cells and resulting in markers of gene expression typical of "younger" (early passage) cells without the hallmarks of malignant transformation. It is now possible to explore the fundamental cellular mechanisms underlying human aging, clarifying the role played by replicative senescence. By implication, we may soon be able to determine the extent to which the major causes of death and disability in aging populations in developed countries—cancer, atherosclerosis, osteoarthritis, macular degeneration, and Alzheimer dementia—are attributable to such fundamental mechanisms. If they are amenable to prevention or treatment by alteration of cellular senescence, the clinical implications have few historic precedents.

---

In 2010, a group of researchers at Harvard was able to turn telomerase on and off in a group of mice. This was groundbreaking research and made headlines around the world.[4]

When the researchers suspended the mice's ability to produce telomerase, the mice aged more quickly than the control group. The mice remained in this condition until adulthood, and they became much less fertile and experienced other age-related conditions, such as diabetes and osteoporosis. Some of them died prematurely.

But when the researchers reactivated the mice's ability to produce the enzyme by adding the chemical 4-OHT for a month, an amazing thing happened. After another month's time, the mice experienced "a near 'Ponce de Leon' effect," wrote one of the researchers, in reference to the explorer who searched for the mythical Fountain of Youth in the fifteenth century.

Shriveled testes returned to normal size and the mice became fertile once again. Degenerated organs—brain, spleen, liver, intestines—recuperated from their advanced stages of aging. Their brains grew in size, with neural progenitor cells (which produce new neurons and support brain cells) functioning normally again. The mice's fur even changed from an aging gray back to black.

Can the normal pace of aging in human cells be slowed down and reawakened by telomerase?

Although this research certainly supports the idea that telomerase could be a serious anti-aging intervention, how we translate mice research into human research is another giant leap of science. Many researchers warn that telomere "rejuvenation" is potentially very dangerous because it could stimulate the growth of tumors, a.k.a. cancer.

Also, we can't eat or inject telomerase, because the enzyme only works inside the cell nucleus. The only way to kick-start an increase in telomerase is to prompt our cells to produce it. Some branded supplement formulations claim to have an effect on actually halting or reversing telomere length, but these are not proven to work—at least not yet. No solid research study involving humans warrants the expense of $300 or more a month for such branded products. More important, we don't know yet if any "telomerase activation" products could unintentionally stimulate cancer.

At this moment, you may be thinking: "Here's another naysayer doctor trashing supplements."

To the contrary. I've been working with a naturopathic doctor as part of my team for more than ten years. Dr. Shelley Burns and I agree there are supplements, such as glutathione, Vitamin D, and fatty acids, that promote the health of telomeres, many of which Dr. Elizabeth Blackburn herself advocates. You'll read about these in Chapter 6.

I'll end this chapter with a quick "Genetics 101" so that you can read the condensed version of the Nobel Prize–winning research and appreciate just how groundbreaking the scientists' work was to advance the science of aging—and the prospect of billions of people some day living longer, healthier lives.

## Genetics 101

Much like how countries are shaped and driven by their citizens, our bodies are formed and maintained by our cells. Different types of cells fulfill specific functions, and they work with one another to create a stable environment to allow individual cells and our entire body to thrive. Maintaining this ideal stable state, or homeostasis, is not an easy job. Outside factors and stimulants constantly test and stress our bodies. Changing weather (such as extreme cold and hot temperatures), caffeinated drinks, the food we ingest, and germs are just some factors that can upset this equilibrium. To respond quickly and efficiently to these changes before permanent damage occurs, our 10 trillion cells serve specific functions in different tissue sectors, which band together to form *organ*izations that we call organs. The functionality of one is dependent on the other, and the combined and constant effort of all our cells is what it takes to stay alive.

Each of us begins life as a single cell. That one cell splits into two cells, those two cells become four, four to eight, and so forth. Within the first cell, and in every cell after that, is a set of deoxyribonucleic acid (DNA), a full collection of genetic coding inherited from our parents. Each time a cell splits, the DNA is copied and passed on to the new cells, called daughter cells, that form.

Our DNA "encyclopedia" is organized into twenty-three pairs of chromosomal "volumes." Each holds thousands of gene "chapters" with detailed instructions to build complex proteins. DNA is coded in nucleotide bases, like a genetic alphabet made up of only four characters: A for adenine, C for cytosine, G for guanine, and T for thymine. Each set of three nucleotide bases forms a "word" representing one of twenty amino acids. These are compounds that arrange themselves into a long chain that coils and bends to form functional units called proteins.

Proteins come in various shapes and sizes. They can have anywhere between 31 and 27,000 amino acids. Collectively, proteins make up approximately 45 per cent of our bodies. Different amino acids interact differently with one another within the protein: some attract each other to form strong bonds; others repel one another. The properties of each amino acid determine how the protein is wound up, and the shape and reactivity of the protein determine its function.

Your body synthesizes these proteins, which are responsible for almost all of its processes, including cell organization, cell signaling, cell repair, waste cleanup, and synthesizing other proteins. Enzymes, hormones, and some of your immune molecules are all made of proteins, and proteins act as the body's main energy source.

Everyone has a slightly different set of DNA. This means everyone produces a slightly different set of proteins, giving each of us one-of-a-kind qualities. The difference is small, approximately 0.1 per cent, but it's enough to give us our distinguishing features, such as the color of our eyes, skin, and hair and the strength of our muscles and bones.

Most cells in our bodies, in fact all except our brain cells, are completely and repeatedly replaced by new ones, some more frequently than others. Our skin turns over completely every three months, but other organ cells do so every sixteen years. When we're young, new cells divide very frequently and, as a result, we grow in size. As we age, and because our cells have undergone their share of divisions, this process slows down.

We all understand that when we copy something in a photocopier, the copy is the same as the original. It is an exact replica. However, the replication of our DNA is a dynamic thing and mistakes happen. In fact, sometimes the replication process gets lost in the translation, and we can end up with a defective protein. Thankfully, there are numerous mechanisms that prevent copying errors like this from happening in your body. One mechanism is your telomeres, the tips of your DNA, and the other is the enzyme telomerase, which forms telomeres.

In the 1930s, two scientists, Herman Joseph Muller and Barbara McClintock, observed that the ends of chromosomes (what we now call telomeres) seemed to prevent chromosomes from sticking to each other during cell division. A few decades later, scientists came to understand more about this process and realized that chromosomes should, in theory, shorten every time a cell divides (because for every one of two DNA strands, a section of the strand can't be copied). But chromosomes don't shorten at each division. Why? The answer is telomerase. For this discovery, Muller won the Nobel Prize in Medicine in 1946, and McClintock won it in 1983.

Later, Dr. Elizabeth Blackburn was able to map out a section of single-stranded DNA known as the telomere portion of the chromosome. Part of her Nobel Prize–winning research with Dr. Jack Szostak showed that this sequence protected DNA from degrading. Years later, Dr. Blackburn and her then-graduate student Dr. Carol Greider demonstrated that telomerase also has the ability to delay the senescence, or automatic cell death, of human cells.

FURTHER INFORMATION

To read more about the Nobel Prize discovery, visit nobelprize.org/nobel_prizes/medicine/laureates/2009/press.html.

Now that you have an understanding of what telomeres are, I will turn to how this will affect you in your everyday life. In anticipation of the program detailed in the second part of the book, the next chapter lays out an inventory of the various ways we have of keeping track of the aging cycle in our bodies.

# 4

# Know Your Biomarkers and the Aging Processes

## Biomarker Basics

In the first three chapters, I introduced you to telomeres, which determine your lifespan. When people die, they have generally not used up all of the telomeres on their chromosomes. That means their telomeres have not reached terminal lengths, which signals cell death. In fact, many scientists believe we are technically programmed for self-repair and longevity rather than being programmed to age and die. These scientists believe that the existence of telomerase in animals like humans means there is a capacity to repair indefinitely and live for a very long time. Why telomerase is not turned on all the time in humans is the subject of current scientific research and speculation. Remember that the telomerase enzyme exists to repair and re-lengthen our telomeres. As I lay out the Peak Health Program over the rest of the book, I explain how this enzyme "activates" when we are doing positive things with our bodies for the purpose of staying alive.

DEFINITION

**Anti-Aging Medicine** "The speciality of medicine that seeks to slow age-related disease process to improve both the quality and quantity of life."[1] Vincent C. Giampapa

These telomeres, the "aglets" of our chromosomes, also determine the stability and structural integrity of our DNA. When telomeres are compromised, errors occur in our DNA replication. This, in turn, compromises our ability to produce proteins and enzymes, which ultimately leads to organ damage and disease.

There are other factors that age us before our time and cause disease before it's biologically time for us to die. When we injure or harm ourselves due to our lifestyle, accidents, or infections, our bodies go into a hyper-healing mode, repairing our damaged tissues by replicating our DNA. As we repair our bodies, we use up our telomere length.

Once you've calculated your biological age and your rate of aging by measuring the length of your telomeres, you then need to identify which processes are causing them premature damage. We quantify these processes through biomarkers. These are measurable indicators that we know will change as we age and that can be tracked over time. Biomarkers can also be manipulated through our diet and lifestyle, and by taking supplements and medications. As with telomeres themselves, if we can improve them, we can slow down the processes of aging.

DEFINITION

**Biomarker** The National Institutes of Health defines a biomarker as "a characteristic that is objectively measured and evaluated as an indicator of normal biologic processes, pathogenic processes, or pharmacologic responses to a therapeutic intervention."[2] Stedman's *American Heritage Medical Dictionary* defines it more simply: "A specific physical trait used to measure or indicate the effects or progress of a disease, illness, or condition."

What's more important than the official definitions is knowing what biomarkers mean for your health. In short, biomarkers can identify the aging processes at play in your body. In this chapter, I review what those biomarkers are and how they shorten or lengthen your telomeres.

---

## Types of Biomarkers

There are ten biomarkers for aging. Three of these are traditional functional biomarkers, and seven are cellular biomarkers that track your body's aging.

### TRADITIONAL FUNCTIONAL BIOMARKERS

*Cardiac Function* When you are sitting quietly, the volume of blood pumped by your heart each minute decreases over time—so the older you get, the less blood moves through your body. This is called decreased resting cardiac output.

*Lung Function—Forced Vital Capacity* Your lungs become less elastic and less effective as you age too. This means their ability to take in and expel a certain volume of air weakens over time—this is called a change in forced vital capacity. A mortality study that lasted two decades, where patients were tested every six months, confirmed this aging biomarker.[3] The cause of this effect could be a loss of muscle power, or a weaker diaphragm or muscles in the chest wall.

*Kidney Function* The older you get, the slower your kidneys work to filter waste from your body. This trend can be observed in your mid-forties. Creatinine is a chemical waste molecule produced by your body. Elevated levels indicate that your kidneys are not flushing the chemical from your body effectively. A creatinine test can measure your renal, or kidney, function.

### CELLULAR BIOMARKERS

*Body Fat, and Muscle Mass and Function* As we age, we tend to store more body fat. Doctors use skin calipers to measure the thickness of a fold of skin and fat and compare it to previous measurements. Aging is also associated with a decrease in muscle mass, meaning that our muscles become smaller and weaker over time. We can test our muscle strength using traditional fitness tests, such as hand-grip strength and physical endurance.

*Bone Mass* Our bone density decreases as we age, and bones become more prone to fractures, breaks, and skeletal collapse. This is especially true with older women. How many older people do you know who have broken a hip? This bone mass decrease is called osteopenia (early bone loss) or osteoporosis (medically worrisome bone loss) and can be calculated by performing a radioisotope bone scan.

*Fingernail Growth* Even though fingernails grow continuously through our lifetime (sometimes even after our death), scientists have discovered that how fast our nails grow slows over time—by as much as 50 per cent over our lifetime.

*Neurological Function* Decreases in our neurological, or brain, function, such as hearing, sight, reaction time, and memory, may also be noticed as we grow older. In fact, some people talk about their memory loss as much as the weather. All of these brain functions can be easily measured using computerized testing, which is available through many clinics.

*Skin Changes* Our skin's elasticity decreases over time. You may have noticed that your skin doesn't return to its natural state quite as fast as it used to following a gentle pinch of the skin on the back of your hand. This loosening is called turgor and it's also a measure of dehydration.

*Hormone Levels* Hormones are a hot subject with the media these days as scientists, doctors, naturopaths, dietitians, and others make connections between our hormone levels and our overall health and wellness. We do know that many of our hormone levels fall over time. Some levels decrease linearly, or in a predictable way—such as dehydroepiandrosterone (DHEA) and melatonin. But other hormones—such as testosterone and estrogen—rise and fall more irregularly over our lifetime. You can read more about hormone-replacement therapy in Chapter 7.

*Immunological System* Our immunological responses change over time as well. An immunological response is our body's reaction to something it detects in the blood—like a virus or allergen—something that is foreign and potentially harmful. In reaction to this detection, the body produces antibodies to attack and neutralize the threat. As we age, our immune responses decrease, which explains why older people are more likely to become ill and recover more slowly from an infection.

We can determine our body's immunological strength through blood tests, such as measuring different types of white blood cells and antibodies.

### Controlling Biomarkers

Once you understand your biomarkers, you can then control them. There are three main ways to gain that control:

1. Control your diet, activities, exercise, and sleep—and reduce your emotional stress and your exposure to toxins, pollution, and radiation.

2. Slow down the processes of aging by reducing glycation, inflammation, oxidation, and more.
3. Improve the odds of DNA repair minimizing cell mutations and making accurate copies of your DNA.

It might help to compare our DNA to air travel. Each day, around the globe, thousands of airplanes are in the air at one time. At a single large airport, hundreds of airplanes are coming and going, on the ground and lifting into the air. Air traffic controllers the world over keep these airplanes in check, moving them along at the right speed and altitude to minimize congestion and prevent collisions. This system works spectacularly well.

Within an average adult human body, there are an estimated 100 trillion cells and billions of miles of DNA—or approximately five feet of DNA in each cell. Under peak conditions, our DNA makes perfect reproductions of itself. This process helps the body to grow and repair itself. Each day, however, our body is bombarded by environmental factors that damage our DNA. What a feat it is to keep us moving even when we don't feed our bodies properly, become sedentary and unfit, and give ourselves little rest for recovery. It's truly a wonder what we take for granted every day. Just like all those planes in the air, your body needs a traffic controller to improve your odds of DNA repair. That controller, of course, is you.

## Medical Checkups and the Advance of Medicine

Biomarkers are a helpful window that can reveal what's going on inside you. Before you read this chapter, was the word *biomarker* new to you? I wouldn't be surprised if it was. A few more questions will reveal why I think this way.

- Over the past fifteen years, how much has your annual physical checkup focused on the biomarkers of aging?

- Have you been given specific information that could prevent you from developing cancer, dementia, or diabetes?
- What's changed in terms of the diagnostic tests you've received to help you quantify and learn about your body?
- What new lab work and imaging tests have you been offered? Nothing really new? Again, I'm not surprised.

The checkup I learned to perform in medical school twenty-five years ago is pretty much the same one that medical students learn today. This, despite all the scientific discoveries that can help determine your wellness—and your sickness. We are still taught to use our hands and eyes to evaluate a patient's health. We use stethoscopes, tapping hammers, tuning forks, and scales. We sometimes order blood and urine tests and occasionally X-rays, mammograms, and bone density tests.

But that's it.

The trouble is, this examination isn't effective in catching the warning signs of the three big killers: heart attack, stroke, and cancer. Nor does it enable doctors to diagnose diabetes early enough to prevent further damage to your body. And if we want to learn about your rate of aging? Forget it.

This isn't just one doctor saying this. The highly regarded medical resource the *Cochrane Database of Systematic Reviews* published a paper in 2012 titled "General Health Checks in Adults for Reducing Morbidity and Mortality from Disease." It concluded that checkups are, in essence, useless. The researchers noted: "With the large number of participants and deaths included, the long follow-up periods used, and considering that cardiovascular and cancer mortality were not reduced, general health checks are unlikely to be beneficial."[4]

The reason I know this study makes sense is that today's checkups don't adopt predictive biomarkers for diseases. Too many physicians focus on "standard of care," which is a minimum measurement established by committees of specialists as guidelines for their physician colleagues. Such guidelines provide a safety net for physicians to perform just those tests as a way of managing costs and satisfying a minimum agreed-upon standard of testing.

But many doctors, including myself, far prefer to offer our patients "optimal care." This means that sound scientific tests should also be used, whether or not they're "free" via a public health care system, or available through private insurance or as self-pay.

What really determines if a new test or technology is used in either the Canadian or American health care systems isn't the effectiveness of the test, but politics and dollars. We've all heard that many tests are no longer covered by health insurance for the so-called average-risk person. Why? Statistically, the costs of the test will outweigh the benefits.

No one ever says that these tests don't work. Rather, the system can't afford to offer the test for everyone and is willing to sacrifice a few—and miss the diagnosis entirely. Too bad for those people who would have benefited from the detection of a tumor. And tragic for you if you're one of "the few."

Ethically, for me, this is unacceptable.

Today, patients, or rather health consumers, can venture beyond standard care and pay for their own medical tests. You can easily get a PSA test to detect prostate cancer, a CA-125 test for ovarian cancer, or a test that checks your nutrient markers to detect vitamin D deficiency. But these detectors are only the most basic of health predictors—table stakes, if you will.

Indeed, there are many more effective and precise tests available today. Many of them are products of the technology revolution.

The first BlackBerry was introduced in 1998 and the iPhone launched in 2007. Ten years ago, Facebook, Twitter, and LinkedIn didn't even exist. Most of us change our mobile device every eighteen months to two years. Today, our smartphones can track our heart rate and our blood pressure through wearable and portable monitoring devices. Even our glucose levels can be measured through a contact lens (Google's invention). Soon, our wristwatches will become dashboards for the state of our lives and tell us when we need to drink more water or get out of the sun.

So why hasn't the North American physical examination kept up with the rate of change in medicine? Let's look at a few factors.

1. The health consumer is unaware of what is available to them. Too often, medical health stories in the media focus on the next cure or treatment rather than on strategies that are proven and exist today to prevent disease.
2. Too much medical school training for doctors is focused on disease care. Very little has been done to advance preventive care. True, there is a gradual shift to focusing on anti-aging and preventive medicine. But it's painfully slow.
3. There's real inertia in demanding that new tests be covered by public and private health insurance. I suspect this is due to the inability of bureaucrats to adjudicate in a knowledgeable way and approve the value of new diagnostics in a timely manner as these tests come to market.
4. Patients are reluctant to pay for additional tests.
5. Assuming physicians understand the value of a test, they feel uneasy about telling patients that a test is important and that they have to pay for it.

## New Technology for Old and Growing Problems

One factor in this reluctance is that some decision-makers aren't convinced of the effectiveness of new medical tests. If that's the case, they would be tragically misinformed. There are plenty of new medical tests that doctors can perform to detect or give advance warning of serious and sometimes fatal conditions. Put simply, these tests are much more effective than the ones we use today.

For example, GTA-446 is a colon cancer risk assessment marker. It measures a specific type of long-chain fatty acid, which is detected in blood serum. If you have a low level of this nutrient, your risk of getting cancer is as much as three times higher than that of those who have a normal value. This marker is a more powerful screening tool than just relying on detecting blood in a stool through a fecal occult blood test, which doctors use today to determine if you need a colonoscopy. Just a simple hemorrhoid bleed can cause today's test to be positive, which then leads to a colonoscopy, a costly and unpleasant procedure.

Most bureaucrats think these new tests, like GTA-446, would cost the health care system scarce new dollars. On that point, they would be right—although the cost for many tests would plummet if they were performed on millions of patients rather than on thousands or even just hundreds.

What determines if the cost of a GTA-446 test is covered should be based on whether it can reduce the number of colonoscopies being performed now. In the United States, it turns out that the number of colonoscopies rose sharply after they were fully covered by Medicare and private insurance. In Canada, the experience is the same. Imagine the politics of removing the colonoscopy revenue from the hospital and physician side of the ledger (at $900 to $1,500 per procedure) and replacing it with a $75 blood test.

So new tests can also dramatically *reduce* overall health care costs as well. For example, new tests can keep patients out of emergency rooms by detecting a condition, such as a tumor or a heart condition, before it has time to show itself with obvious symptoms.

Consider heart attacks. I believe they are now totally preventable. So does Dr. Sanjay Gupta, chief medical correspondent for CNN. He made his case well in the CNN documentary *The Last Heart Attack*.[5] Using a combination of predictive biomarkers and a computerized tomography (CT) scan to perform a virtual angiogram, he demonstrated that we have the technology today to predict—and therefore prevent—heart attacks in all industrialized nations.

Technological innovations of the hand-held variety also have a crucial preventive aspect. There's a reason smartphones are called smart. There are hundreds, if not thousands, of new apps that can help detect health problems or irregularities. Today's smartphones can assess our sleep patterns, track our activity levels, and monitor how much food we eat each day. Other apps can detect life-threatening symptoms, such as low or high blood sugar, irregular heartbeats, or a change in the oxygen level in your blood. These apps will not just notify you, they'll also send a message to your physician. Google Glass may soon be able to measure your brain waves and detect the onset of drowsiness or, more seriously, seizures. Who can even guess what devices will be able to do for us in another five years? One thing we do know, however: the information they provide will be more accurate than what we receive today.

> I think the biggest innovations of the twenty-first century will be the intersection of biology and technology. A new era is beginning.
>
> WALTER ISAACSON, *Steve Jobs*

## Modifiable and Non-Modifiable Biomarkers

Biomarkers can be modifiable or non-modifiable. Modifiable biomarkers change when their environment changes. For example, our muscle mass and aerobic lung and heart capacity can improve if we change our exercise regimen and our diet.

Non-modifiable biomarkers are genetic characteristics, such as our height and bone length, and they can't be changed by diet, medication, or lifestyle. Although non-modifiable and modifiable biomarkers are one way to classify biomarkers, a more useful distinction is to classify biomarkers as subjective and objective.

As more and more laboratory tests become available, we can become more precise about quantifying a marker.

Historically, doctors have had to rely on an interview or questionnaire to obtain a family and personal medical history. They'll ask you if your grandparents, parents, and siblings have experienced heart trouble, suffer from diabetes, or have survived cancer. As well, they ask for the cause of death of your extended family members. The answers to these questions are known in data terms as subjective, or blunt, biomarkers. And although your responses may give a sense of what your genetic predispositions or disease risks may be, the answers can't easily be quantified for future comparison.

But doctors today can be much more objective and accurate by being able to map a patient's entire human genome.

Every so often, we hear that a birth control pill increases the risk of developing a blood clot that can lead to a pulmonary embolism (a blood clot traveling into the lung) or stroke and even sudden death. But our DNA can easily tell us that some people are born with a genetic defect that can cause them to excessively clot in certain circumstances, such as when they're sitting down for an extensive period of time. This is why doctors will often tell their patients with this condition to walk up and down the aisle during a flight, or break up a long car drive with a walk to keep their blood moving. We also know that ingesting hormones can increase some people's risk of clotting. Both of these concerns can be addressed to some degree with a simple blood or saliva test for factor V Leiden mutation to identify if you carry a risk for developing a deep vein thrombosis (blood clot).

DEFINITION

**Factor V Leiden** The incidence for a factor V Leiden mutation in the population can be as high as 5 per cent depending on your ethnic background. Having one gene for this mutation carries an increased risk of 300 to 800 per cent, and carrying two genes can increase the risk by up to 80 times. Naturally, if you carry this gene, you should not take hormones and you should be on high alert for developing a clot.

### Genetic Testing

I can attest to the power of having a warning sign from genetic testing. A patient of mine knew her father had died from a blood clot in the lungs. She proved positive for having a genetic mutation for factor V Leiden. The patient noticed a mild swelling in her ankle after shopping one day, and we agreed to test the area for abnormalities. The test showed that a blood clot was forming in her lower leg. This was discovered well before pain ensued or, worse still, shortness of breath (which could mean that the clot had traveled to the lungs). She was immediately treated and her symptoms disappeared. In 2013, actress Angelina Jolie had a double mastectomy because she discovered she carried a genetic mutation known as BRCA, which increased her risk for breast cancer to almost 87 per cent.

These are just two examples of the power of genetic testing. Understanding genetics and being able to alter our genetic blueprint is one of the most powerful ways to prevent the onset of disease.

SCIENCE STAT

**Pharmacogenetics** Today, with pharmacogenetics, we can understand how our bodies break down medications. It is the most objective way to prescribe a medication. We know which blood pressure medications will work better for you and what type of pain medications will give you relief. More important, we can predetermine which drugs will cause unpleasant side effects or not work at all at certain doses. So you don't need to become a drug experiment waiting to happen in the medical office or in the emergency department. Indeed, you no longer have to be your own lab rat!

One of my patients has a history of an irregular heart rate. He was in and out of the ER because his doctors had trouble controlling it with medications. The most common class of drugs to combat this is beta-blockers, which slow down your heart rate. He was taking many different beta-blockers. None of them worked, and eventually his cardiologist decided to operate on his heart. Thankfully, the procedure worked. When I met him years later, we decided to do a genetics panel. To my surprise (or perhaps not), the testing showed that his body broke down beta-blockers ineffectively and the beta-blockers were not available in their active form to benefit him. In other words, he didn't process this particular medication effectively. Had he known this when he developed his heart condition, he would have saved months of trial and error taking medications that didn't work to regulate his heart rate. Not only did those medications make him feel unwell, but it was dangerous to have a continual irregular heart rhythm, which increased his risk of having a stroke.

Although taking something like ibuprofen (Advil™) for regular aches and pains doesn't seem like a life-threatening concern, it can be if you can't process the drug properly, and especially if you're taking ibuprofen in combination with another treatment. Here too you can take advantage of understanding what doctors call your pharmacogenetics to learn if a drug like ibuprofen (which can improve the quality of your life) really works for you.

SCIENCE STAT

**Nutrigenomics** The idea behind nutrigenomics is to eat according to your genetic profile. It's a scientific fact that we all respond differently to the foods and drinks we consume. But today, you can determine the effect of salt on your blood pressure; whether you are lactose intolerant; how well you can absorb certain nutrients, such as folic acid, vitamin C, and omega fatty acids from the food you consume; and even whether caffeine is "good" for you—a popular question!

In the past five years, genetics companies like Pathway Genomics and 23andMe have offered health consumers population genetics testing. This is a statistical calculation of your genetic risk based on a group of naturally occurring genetic markers, or SNPs (single-nucleotide polymorphisms). These SNPs have been found to increase the risk of certain disease conditions, especially in diseases such as diabetes, Alzheimer's disease, macular degeneration (the leading cause of severe vision loss in people over 60), atrial fibrillation (a kind of cardiac arrhythmia), and certain types of cancer. Geneticists at these companies bundle a group of SNPs into a condition and provide a statistical estimate based on how many of the SNPs you possess. If you score higher than the average population, you should probably keep an eye on the condition and take steps to screen yourself more often. Currently I suggest patients revisit their genetics profile every few years and retest; a genetics panel becomes more accurate and precise as more research is done. I believe we will begin to move toward an era of exome sequencing which is short of whole genome sequencing, making it less costly. Exome sequencing consists of selecting only the subset of DNA that encodes proteins (known as exons), which allow us to identify genetic mutations that are responsible for many disease states.

I liken this type of genetics assessment to the way stocks in a mutual fund are selected to maximize their return. A portfolio manager picks a group of companies—say, in the mining or pharmaceutical sector—for the fund portfolio. Higher-risk equities might be included in the mix. Bundling a group of companies together improves the overall odds of making money in certain sectors of the economy. In the genetics world, the idea of bundling a group of SNPs or exomes together is to improve the odds of measuring your personal risk of disease. Although it is beneficial to see groups of potentially higher-risk equities perform well in your retirement savings, the opposite is true in the world of assessing disease risk genetics. You want to de-risk your health as much as you can. My preference is that your genetics portfolio reveals no risky SNPs or exomes in your testing, so you don't have a higher risk of developing a disease based on what we know today in genetics research.

Not everyone supports this kind of genetic testing. Naysayers of genetic testing believe that although the techniques of identifying the SNPs and exomes are accurate, they don't specify which SNPs are the key genes that contribute to the risk of a disease condition.

My rebuttal is: when do we decide that a measure of accuracy is good enough? Science is always evolving. Had Madame Marie Curie not pushed forward the notion that the X-ray was an important tool in assessing injury in the battlefield during World War I, we might not be as far along in our imaging technologies as we are today, with the hugely beneficial insights provided by CT and MRI scanners.

Someone had to question the status quo. The medical world fought her, and in the end, she won the debate and became the first woman to win a Nobel Prize. In fact, she won the Nobel Prize twice.

While many traditionalists may not agree with me that genetics can be modified (via our environment), certainly some biomarkers are totally modifiable by our lifestyles. In Chapter 2, I highlighted the three main processes of aging—namely, glycation, inflammation, and oxidation. Some biomarkers can give us an indication of how "rusty and inflamed" you are really becoming. These processes all affect the rate of your telomere shortening.

### Know the Biomarkers of Glycation and Diabetes

Glycation occurs when cross-links (a criss-crossing) of proteins develop at the cellular level. This can cause your organs and DNA to function poorly.

This process directly affects your gene expression and protein production. Glycation is caused by poorly controlled glucose levels in your bloodstream, by poor diet, by insulin surges, and by insulin receptor insensitivity. You might recognize these medical words in definitions for diabetes, prediabetes, and insulin resistance.

Biomarkers that can measure glycation are made up of many diabetes risk markers. These include the commonly tested fasting blood sugar (glucose) level. Although constantly high blood glucose levels indicate diabetes, I believe that taking such snapshots in time can't really determine if you're becoming diabetic because this marker becomes elevated only when your body systems are broken and can no longer cope with the sugar load. Sadly, by then it's too late. Using a car tire analogy, why wait until all the air is out of the tire to do something about it? Once there is a hole in a tire, it is defective forever. The same is true with the diagnosis of diabetes. Once you are a diabetic, you are always a diabetic. At best, you will perhaps be a diabetic who is in "good control" of your high blood glucose levels, through diet and medications.

I believe that hemoglobin A1C (HgbA1C) is a more effective biomarker than a fasting blood sugar measurement. I've been using this marker as a guide to show my patients they're starting to show the signs of diabetes with abnormal HgbA1C levels. It measures the three-month control of your blood sugar levels in your body. In this test, we are measuring the degree of sugar coating on your hemoglobin (red blood cells), which swims in your body for a three-month lifespan. Finally, as of 2013, both the American and Canadian Diabetes Associations advocate the use of HgbA1C as the better standard to diagnose diabetes.

There is no doubt that diabetes is associated with the presence of short telomeres,[6] which may be why so many chronic conditions are linked to diabetes. One of these conditions is vascular injury leading to organ damage of the heart, brain, kidneys, and eyes.

Other high-potential tests to measure your diabetes risk include measuring your insulin and adiponectin levels. Both are very important hormones that regulate blood sugar levels. Certainly, these biomarkers can predict insulin resistance, a condition where a person's cells have difficulty responding to the normal actions of the insulin hormone.

Researchers have shown a relationship between insulin resistance and diabetes, and therefore to telomere shortening. This relationship is especially strong when it coincides with an increase in your body mass index (BMI).[7] In other words, the more weight you gain, the more damage is done to your cells and organs. Why? When insulin resistance increases, oxidative stress also rises—and oxidative stress, as we now know, speeds up the degradation of telomeres.

DEFINITION

**Body Mass Index** BMI is a measure of your body fat based on height and weight. The BMI categories are:

| | | |
|---|---|---|
| UNDERWEIGHT | = | less than 18.5 |
| NORMAL WEIGHT | = | 18.5–24.9 |
| OVERWEIGHT | = | 25–29.9 |
| OBESE | = | 30–34.9 |
| MORBIDLY OBESE | = | 35 and greater |

To calculate BMI, we use the equation of mass (kg) ÷ height $(m)^2$. For example, a man who weighs 210 lb (95 kg) and is 5'10" (1.78 m) in height would be calculated as $95 \div 2.13^2$, which equals a BMI of 30, which would make him obese. Do you know your BMI? If not, figure it out. There are numerous BMI calculators on the Internet.

## OBESITY

Obesity is by far the most serious public health concern of the modern industrialized world. The epidemic rise in obesity and diabetes means we are rapidly aging prematurely as a society. Children are diagnosed with diabetes at a younger age. Most alarming? Many children are now expected to die before their parents: not because of war or infection, but from complications of diabetes—beginning with cardiovascular disease and heart attack (read more about this in Chapter 5).

I applaud the American Medical Association for declaring obesity a disease in 2013. The announcement caused a stir across the nation, and not just in the medical world. Comic Bill Maher quipped, "Wow, people are already taking advantage of this. They're calling in sick with a case of the 'fats.'" But obesity is hardly a laughing matter—and it takes a real toll on our personal health and on productivity in the workplace.

Many people still don't realize that obesity significantly increases their risk of at least eighteen chronic diseases. Some of these can be fatal. The three big killers—heart disease and stroke, diabetes, and cancer—top the list, followed by other conditions like high blood pressure, arthritis, and liver and gall bladder disease.

Medical studies reveal that obesity speeds up the aging process. It has been demonstrated that when a person gains weight, their telomeres shorten at a faster rate. Indeed, this research revealed an inverse relationship between a person's telomere length and their BMI and hip circumference. In other words, the higher your BMI and the wider your hips, the shorter your telomeres will be.[8]

But there's hope for us yet. Researchers have discovered that obese children do not have shorter telomeres than children with healthy body mass indexes, so this is a result of nurture, not nature.[9] This discovery reinforces the importance of addressing childhood obesity at a young age so we can reduce the long-term risks for chronic illness associated with age and obesity.

If you have visceral (abdominal) fat outside your body, it means there's fat on the inside of your body as well—because there is fat inside our organs too, which impairs their function. For example, ultrasound imaging can pick up visceral fat in an organ, what we often call fatty liver. People with fatty livers often have impaired cholesterol metabolism, and we know that cholesterol has effects on our vascular health.

SCIENCE STAT

**Cholesterol** Imagine a smooth plastic pipe. Now pour lard through it. Will it stick? No. But what if you gouge the inside of the tube, scratching it up? Now the lard gets caught in the wall. That's what high levels of sugar are doing in your bloodstream. Sugar gouges the lining of your artery walls. Cholesterol comes along, and sticks to the wall. Add a bit of calcium, and voila! You have plaque—a permanent scab that grows and, in time, causes atherosclerosis (a hardening of the arteries) and later narrows the blood vessel, causing heart attacks and strokes.

We know that cholesterol is a variable contributing to cardiovascular disease. But studies are also now showing that telomere biology affects your cholesterol levels: telomeres shorten more quickly with higher levels of low-density lipoprotein (LDL, or "bad") cholesterol and they shorten more slowly with higher levels of high-density lipoprotein (HDL, or "good") cholesterol.[10]

Today, measuring cholesterol is as common as measuring blood pressure. But it's not enough to just learn the total and estimated levels of "good" (HDL) and "bad" (LDL) cholesterol. Standard tests calculate these values. We need to go deeper. We need to study all the subparticles of cholesterol. Certain subsets of LDL are more atherogenic (plaque-building) than others, and some HDL subtypes are more helpful in reducing this risk, such as alpha-1. The level of triglycerides (a type of fat in our bloodstream) also plays an important role in understanding our risk for diabetes. An overload of blood sugar causes the liver to store the sugar as these light-chain fats.

---

## Know the Biomarkers of Inflammation

Inflammation is another cellular response we can blame on proteins that damage target tissues and change both our gene expression and the formation of proteins. What determines the rate of our tissue damage is the imbalance of good anti-inflammation nutrients and bad inflammation proteins that are compounded inside and outside our cells. This inflammatory process is controlled by nutrients known as fatty acids, especially along the cell walls.

What does inflammation look like? Well, say you take a tumble off your bike. What do you see after grazing your skin from the fall? Initially, your skin is red and bleeding. Shortly afterward, the injured area is sticky. This is inflammation. And then a scab forms. When the scab falls off, the skin can often be contracted, thin, and tender. How well the repair process goes will determine the level of scarring. Where there's scarring, there's tissue damage.

The effects of aging due to inflammation depend on your net balance of inflammation proteins to your anti-inflammation nutrients. Just like we have debt-to-equity ratios in the financial world, there is a commonly used ratio of inflammation to anti-inflammation that measures the acid levels in our body (arachidonic acid, which is a polyunsaturated omega-6 fatty acid, and fatty acids, specifically eicosapentaenoic acid, or EPA, an omega-3 fatty acid). There are also other proteins commonly measured to detect inflammation, such as C-reactive protein (CRP) and homocysteine.

Research shows us that high levels of C-reactive protein are associated with shorter telomeres and are considered a cardiovascular risk marker, especially for those with diabetes.[11]

And homocysteine protein, which is produced during the breakdown of methionine and is found in foods containing animal proteins, negatively affects telomere length. High homocysteine levels have also been shown to increase the risk of heart and blood vessel disease. One study of 1,300 participants demonstrated that those with the highest levels of this protein in their blood had shorter telomere lengths. In fact, the shortage in these patients corresponded to approximately six years of chronological aging. Six years![12]

There are other biomarkers of inflammation (prostaglandins, leukotrienes, cytokines, thromboxanes, amyloid, and interleukins), but these are rarely measured.

Add the effects of glycation and inflammation together, and the damage worsens.

**Oxidation Effect**

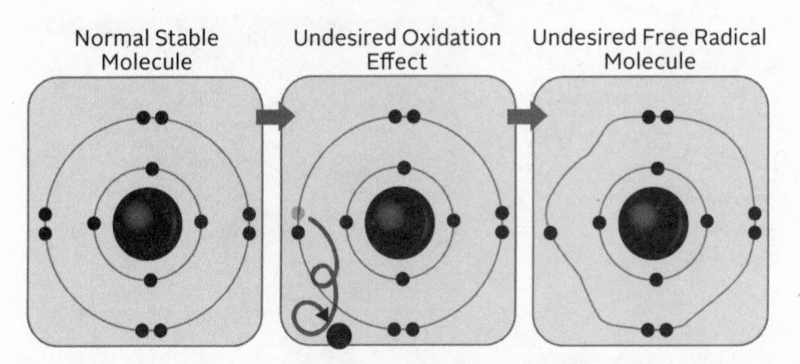


A stable molecule has eight electrons on its outer ring.

When an oxidant attacks a stable molecule, it removes one electron, rendering the molecule "charged" (free radical) and causing tissue damage.

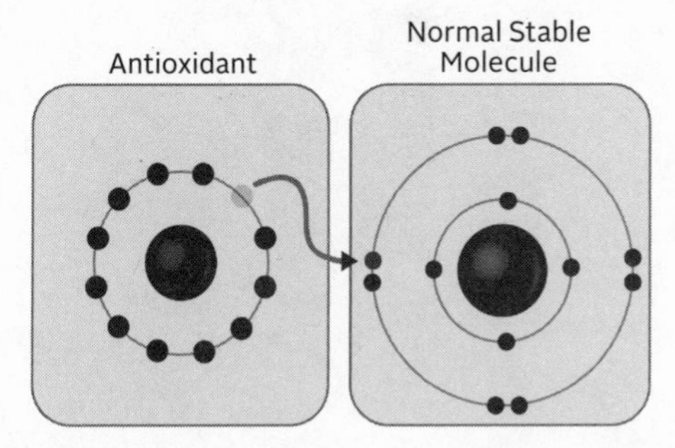


An antioxidant substance donates one electron to the free radical, bringing stability back to the molecule.

## Know the Biomarkers of Free Radical Damage and Oxidation

Free radicals are bad for you. Oxidation, or oxidative stress, is the amount of damage inside and outside of cells created by free radicals. Free radicals are a natural by-product of your body's physiological processes (burning oxygen to create energy for the functioning of cells). They can also enter your body through the environment–from pollution, smog, cigarette smoke, and radiation from sunlight.

**Antioxidant Capacities**

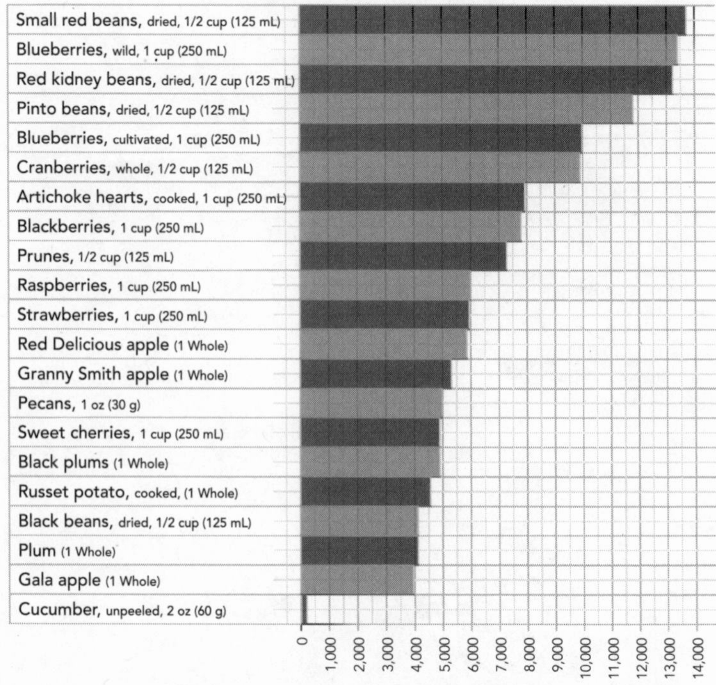


This table shows the antioxidant capacities of different food items. They are listed in order of potency per serving size.

Unfortunately, free radicals negatively affect our genetic structure and cell function by attacking our DNA and damaging cell membranes. This exposes sensitive genetic material that can become prone to mutation and destruction. This process spreads much like rust on metal. The compromised areas intensify and, in the process, weaken your body's ability to neutralize free radicals—leading to chronic diseases such as heart disease, diabetes, cancer, and arthritis, as well as age-related conditions like wrinkles and stiff joints. In other words, the more free radicals you encounter, the more damage done to your body.

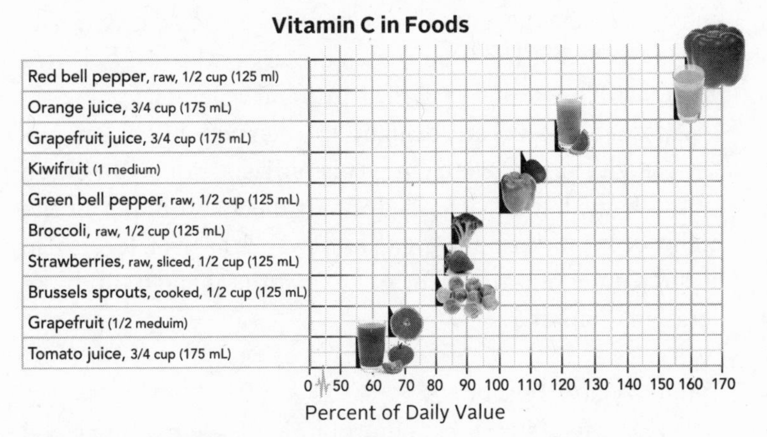


Here are the top ten foods that contain vitamin C, according to the National Institutes of Health.[13]

One way to really "see" oxidative stress at work is to watch what happens to a fresh apple after it's been sliced and left to "age" on your kitchen counter. The oxygen molecules in the air of your home begin to damage the exposed flesh of the fruit, browning it. But sprinkle lemon juice on the slices and the freshness of the fruit is extended. Why? Lemon juice contains the antioxidant vitamin C, and antioxidants can neutralize the effects of free radicals by neutralizing the damaging chemical process.

In the same way, eating or drinking antioxidants or taking them as supplements can decrease the damage done by free radicals and by oxidative stress in your body. Another way to do this is to increase the body's production of natural antioxidants, especially with other healthy lifestyle choices, such as exercising, eating a balanced diet, and minimizing your emotional stress.

As with inflammation, the aging effects of oxidation will be determined in two ways: first, by the net effects of oxidants or free radicals attacking the cell's nuclear DNA, and second, by your ability to prevent damage using extrinsic antioxidants (made outside the body) and intrinsic antioxidants (made inside the body). The best-known dietary antioxidant is vitamin C, which is found in many fruits and vegetables.

Another potent antioxidant is coenzyme Q10, an oil-soluble, vitamin-like enzyme. This nutrient can be produced by the body, but it's absorbed in our diets and found in the highest concentration in organ meats, such as liver. It can also be supplemented in pill form. I explain its benefits in more detail in Chapter 6.

## Immunity and Telomeres

When an infectious threat enters our bodies, our white blood cells immediately begin to proliferate in order to defend against the attack. In other words, they divide rapidly in order to neutralize the threat. Once the infection is under control, most of those white blood cells will die. The ones that remain remind the body how to treat the infection should it return a second time—in essence, providing the body with immunity.

Studies have shown that these white blood cells have shorter telomeres than the original cells because each replication shortens telomeres. Because your immune system depends on the proliferation of white blood cells to treat infection, it's imperative to do all you can to keep your telomeres robust.

It's important to know that immune system cells (which include white blood cells) are unique: they can increase telomerase activity. However, when white blood cells have divided so many times that their telomeres are critically short, it gets harder and harder to activate telomerase, and the body's immunity to infection is compromised.

Our cell membranes are like city walls protecting and guarding the castle that contains our DNA. This membrane controls what comes into our cellular environment and what leaves. Cellular hydration, pH balance, nutrient levels, and repair building blocks (such as the amino acids derived from the proteins we eat) will determine the wall's strength. This is why we are often told to drink lots of water, eat nutritious foods (but not too many acidic ones), and focus on a balanced diet of carbohydrates, proteins, and fats. In fact, amino acids and omega-3 and omega-6 fatty acids are critical nutrients in everyone's diet, and they help keep our cell membranes strong. I explain just how critical these nutrients are in Chapter 5.

### Hormone Balance

The final factor in determining our telomere length is our hormone balance. I've often said that once we pass age 40, we've outlived our evolutionary function, which is to make babies and pass our DNA along to the next generation.

Everything in our hormone soup changes dramatically around the age of 40. Often we focus on the drop of estrogen levels in women and testosterone levels in men. For women, we call this menopause, and for men, andropause. Our muscles, bones, and tissues don't recover as quickly after injury, or even after an exercise workout.

Our physical stamina, energy, and memory begin to wane. What's happening? Simply, our bodies begin to secrete lower amounts of certain hormones needed for reproduction and survival. If this is complicated by an unhealthy lifestyle, which can cause stress and obesity, our other hormones will begin to malfunction as well.

Stress changes our cortisol levels. Cortisol is a hormone, and one of its functions is to help metabolize fat, carbohydrates, and protein. Obesity affects our growth hormone and insulin balance. Lack of sleep lowers our melatonin and serotonin levels. Poor nutrition leaves our bodies with little or nothing to produce critical hormones.

IN THE FIRST four chapters of this book, I have laid out how the science of telomere length provides a crucial way of thinking about our health—and how understanding our lifelines is crucial for taking our health care regime into our own hands. But now is the time for action. Over the second part of the book, I will present the Peak Health Program and provide for you a blueprint for taking your health into your own hands.

TWO

# The Peak Health Program

## An Action Plan

# 5

# Tune Up Your Diet

## Food for Health

I've always said that food is our first-line "drug." Certainly it is our first line of defense in the Peak Health Program. No one can doubt anymore the straight-line link between what we eat and the state of our health. While all of us want to consume foods that promote our health, we also want foods that match our lifestyle, culture, and genetics. Is there a perfect formula for what foods are best for you? Sorry, no. Why? Because your ideal diet may be different from mine. There's no one-size-fits-all solution when it comes to truly personalized nutrition.

So what's the key to finding the *right* healthy diet for you?

You can assess your biomarkers so that you know which facets of the aging processes are going on in your body and what nutrients you're deficient in. Only then can you modify your diet accordingly, and, if need be, add a few key supplements to boost your overall health and the health of your telomeres.

All of that said, I have noticed that my patients share a few very common diet problems. Perhaps you'll find them true for you too. Is the food on your dinner plate balanced nutritionally? Do you ever question the quality of the food you buy? Do you get tired, or "crash," during the day? Could it be you're not getting the right nutrients? Or maybe you have a food sensitivity. Here are four patterns I see all the time.

1. Meals don't have the right balance of carbohydrates, proteins, and fats.
2. The quality of the foods from each food group is questionable.
3. The food is deficient in nutrients.
4. An unknown food allergy, sensitivity, or intolerance is affecting your health (more common than you might think).

SCIENCE STAT

**Is Food Sensitivity a Real Condition?** We all have a friend or family member who limits their diet in some way due to a food sensitivity. Of the many sensitivities—for example, lactose, preservatives, and additives—gluten is one of the most common. In 2012, gastroenterologists finally agreed that gluten sensitivity is a diagnosis separate from conditions such as celiac disease and wheat allergy. The criteria for diagnosing gluten sensitivity go beyond the usual spectrum of gastrointestinal symptoms. Checking for elevated levels of immunoglobulin G (IgG) and immunoglobulin A (IgA)—antibodies to a gluten protein called gliadin—can also help diagnose the problem.

### Getting Your Diet Balancing Act Together

Assuming you are generally healthy and of normal weight, and have no medical issues, conventional wisdom has been to break down your food group proportions into the following percentages:

Carbohydrates: 45%–65%
Proteins: 10%–35%
Fats: 20%–35%

Sound about right? Think again. It did sound good in the 1980s, but it turns out that conventional wisdom was wrong.

This proportion guide came into being in the 1980s for two reasons: cardiovascular disease was on the rise, and research about cholesterol assumed it *caused* heart disease, which was later disproved. In the early 1970s, researchers discovered the LDL ("bad") cholesterol subparticle and later found it to be high in dietary fats. Researchers also saw a correlation between LDL cholesterol and cardiovascular disease. So the logical conclusion was that dietary fats increased cardiovascular disease. Hence, the next step was to reduce the proportion of fats in our diets and substitute them with carbohydrate foods (such as bread and pasta) that were also lower in calories (4 kilocalories per gram compared to 9 kilocalories per gram for fats).

The Heart and Stroke Foundation of Canada, the American Heart Association, the American Stroke Association, and many other medical associations endorsed these guidelines to decrease our fat intake from 40 per cent to 30 per cent and to boost our carbohydrate intake. They all believed that this was the correct way to lower cardiovascular disease. As a result, these guidelines became eating gospel for the next thirty years.

According to a report published by the Centers for Disease Control and Prevention, over a span of three decades (between 1971 and 2000), North Americans' diets changed drastically because the percentage of calories from fat they consumed decreased. This sounds like good news. But it hasn't worked out that way. It turns out that the calories consumed rose by 22 per cent in women and 7 per cent in men over those same thirty years. For women, that is the equivalent of adding 335 calories to your diet every day—or a small hot fudge sundae at Dairy Queen!

How did this shift around cholesterol change the health of North Americans?

- As a society, we are now generally 25 lb (11.5 kg) heavier than we were twenty-five years ago. Shocking!
- Heart disease has not dramatically decreased.
- Obesity and diabetes are at epidemic levels in North America.
- Children are rapidly becoming more obese than their parents and many may even die of complications of diabetes (heart disease and stroke)—*before* their parents die.

So it turns out that medicine got it wrong. To understand why, we need to look at the assumptions these decisions were based on. More than a few assumptions were sadly off the mark.

The first fallacy? That cholesterol is a direct contributor to heart disease. A correlation is not the same as a causality. In other words, just because something's related doesn't mean that the relation involves cause and effect. The problem with the eating strategy in the 1980s was that no one had yet proven that LDL "caused" heart disease.

## Cholesterol: Not a Disease

We were wrong when we started to think of, and "treat," cholesterol as its own disease. Using statin (cholesterol-lowering) prescription medications seemed beneficial because they seemed to slow the progression of heart disease. This observation enhanced the theory that cholesterol caused heart disease. We now know more about what statins do: they may well be acting more effectively to slow down the rate of developing heart disease because of their anti-inflammatory and antioxidant effects. A paper published in 2004 by the University of Iowa Department of Internal Medicine supports this notion. It states: "Several large clinical trials have suggested that the cholesterol-lowering effects of statins may not completely account for the reduced incidence of cardiovascular disease seen in patients receiving statin therapy. A number of recent reports have shown that statins may also have important anti-inflammatory effects, in addition to their effects on plasma lipids." The studies' findings also point to the antioxidant effects of statins, saying they "likely contribute to clinical efficacy in treating cardiovascular disease as well as other chronic conditions associated with increased oxidative stress in humans."[1]

In 2013, the American Heart Association and the American College of Cardiology broadened the pool of people "at risk" for developing heart disease and encouraged them to use an online risk calculator to determine if they should take a statin drug. Both these organizations cited the benefits of this drug class in preventing heart attacks and strokes. As well, they recommended that less focus should be put on the relationship between the drug and the reduction in actual cholesterol levels. The message for us? Take a dose of statin medication—it's good for you, but only if you are at risk.

Anyone serious about telomere health and aging well will learn a lot about fighting inflammation and oxidation during their journey to Peak Health. I believe that statin medications do play a role in reducing plaque formation, but we have to remember that a drug is not a strategy. Drugs are tactics—Band-Aids—which should not be used as a long-term solution. All medications should be used as a stopgap until more sustainable strategies can take hold to resolve the health concern.

Statins are used far too often to support people who should, in fact, use dietary changes as their first-line strategy and long-term solution. Rather than going to a pharmacy and staying on cholesterol medication, people should consult a health professional, such as a nutritionist. This expert can give guidance on what foods should be avoided and increased to reduce cholesterol.

### Sugar: Not Sweet, but Toxic

We know that cholesterol alone can't be blamed for heart disease. The guidelines of the 1980s also suggested carbohydrates were better for us than fats. Wrong again. We now know better. Sugar is a carbohydrate, and there's no doubt about it: sugar is bad for us. In fact, it's so bad that some doctors and scientists have declared a war against it. One such doctor is Dr. Robert Lustig, a pediatric endocrinologist at the University of California. He is one of the first in his field to take serious steps to address the issue of sugar overload in today's society. He's written a number of books on the topic, including *Fat Chance: Beating the Odds Against Sugar, Processed Food, Obesity, and Disease*. You'll want to check out his YouTube lecture "Sugar: The Bitter Truth."[2]

CNN correspondent Dr. Sanjay Gupta is also doing his part to spread awareness about the unhealthy effects of sugar. As a guest journalist on *60 Minutes* in 2012, he asked experts, "Is sugar toxic?" Their answers were telling: they believe in some ways that sugar *has* become a toxin in our society, causing a sharp rise in metabolic syndrome, diabetes, and persistent heart disease.

FURTHER INFORMATION

If you haven't seen the documentary *Fed Up* (2014), do so. It is directed by Oscar winner Laurie David, who produced Al Gore's *An Inconvenient Truth*. It will change the way you eat and think about sugar forever. Check out the movie trailer at fedupmovie.com.

#### OBESITY

Although Canadians are less obese than Americans, the statistics still show that almost one in every three North American adults is obese. According to Statistics Canada, in 2012,

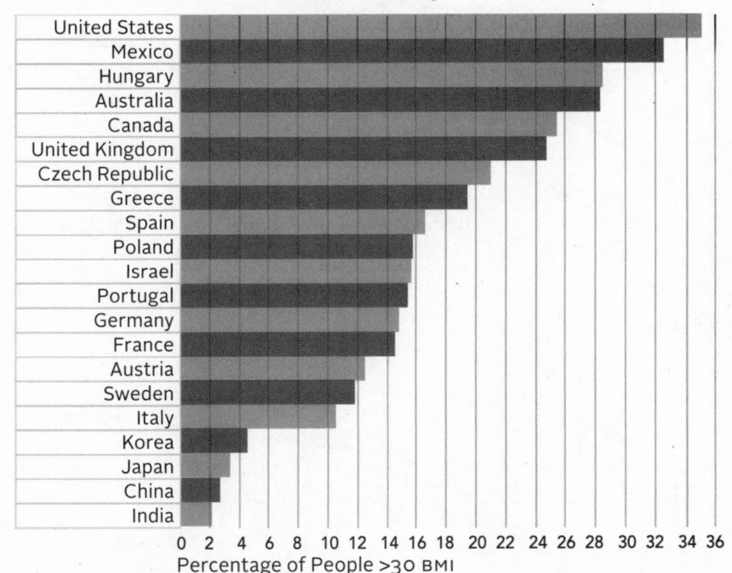


FURTHER INFORMATION

**Obesity: A North American and Industrialized Nations Epidemic**
www.theglobeandmail.com/life/health-and-fitness/health/article13205555.ece/BINARY/w940/if-0713-obesity.jpg

18 per cent of Canadians reported being obese (a body mass index, or BMI, greater than 30) and 34 per cent said they are overweight (BMI between 25 and 30).[3] When Canadians are weighed and measured, however, as in a study by the Canadian Institute for Health Information, the actual figure is 62 per cent—closer to two-thirds of the country.[4] That means almost thirteen million Canadians have an unhealthy weight, and between three million and just over five million Canadian adults are obese, depending on the method of calculation.

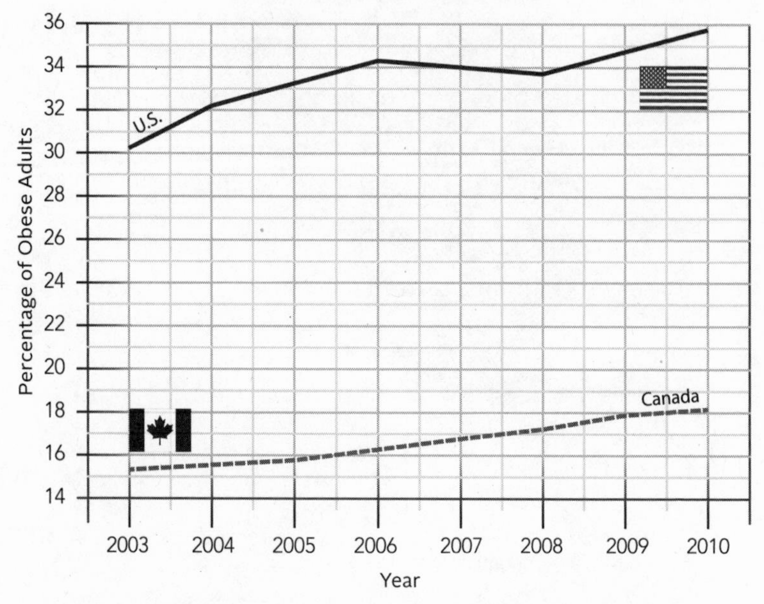


This chart compares the prevalence of obesity among adults in the United States and Canada from 2003 to 2010.

Compare this to American statistics. According to a U.S. Gallup poll in 2012, 26 per cent of American adults reported that they are obese and 36 per cent said they are overweight.[5]

And our children are gaining weight too. Look at the numbers on the following page.

If we stand by and allow this trend to continue, an astounding 70 per cent of us between the ages of 35 and 44 will be obese or overweight in just two decades. This trend also greatly affects the prevalence of type 2 diabetes in children, a condition that once affected only adults. So unless we take action now, today's kids could be the first generation in history to die at a younger age than their parents.

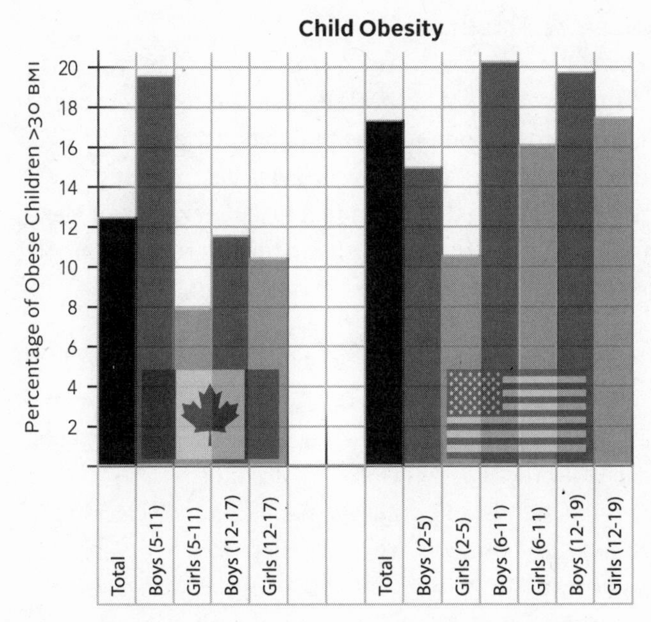


According to the CDC:

- 30% of children 2 to 17 years old are overweight or obese.
- Obesity rates in children have almost tripled in the last 25 years.
- One-third of normal weight 20-year-olds will become overweight within 8 years.
- One-third of overweight 20-year-olds will become obese within 8 years.

---

Today, 12.5 million American children are obese, an additional 13 million are overweight, and more and more of them are developing problems normally found only in adults—high blood pressure, high cholesterol, and type 2 diabetes, which can lead to heart attacks, strokes, blindness, and loss of limbs. If childhood obesity continues to increase, this young generation could be the first in American history to have shorter lives than their parents.
BILL CLINTON, *Giving: How Each of Us Can Change the World*[6]

## THE DANGER OF DIABETES

Those extra pounds are hurting you more than you may think. Being overweight, especially for the long term, can lead to chronic conditions such as high blood pressure, stroke, heart disease, type 2 diabetes, and even death. Being overweight is also associated with an increased incidence of sleep apnea and osteoarthritis. In fact, I don't think there's a single good thing about carrying around excess weight.

### SCIENCE STAT

**By the Numbers** The Canadian Diabetes Association and the American Diabetes Association have published some staggering statistics on the prevalence and future of diabetes.

They estimate that 285 million people worldwide already have diabetes. Seven million people develop diabetes every year, meaning that by 2030, 50 per cent of the population in developed countries will be obese. It's expected that 10 per cent of North Americans will be living with diabetes by the year 2020.

In the United States alone, 25.8 million children and adults—8.3 per cent of the population—have diabetes. Every 19 seconds, one American develops diabetes, and almost half of them don't know it. Almost eighty million Americans—one-quarter of the population—have prediabetes, a medical state that precedes the development of clinically evident diabetes but doesn't reveal any symptoms.

Nine million, or one in four, Canadians have either diabetes or prediabetes; six million of those have prediabetes. Every hour, more than twenty Canadians are diagnosed with diabetes. An estimated 700,000 are undiagnosed.

Here are the American Diabetes Association (AMA) definitions for *diabetes* and *prediabetes* based on blood laboratory test results,[7] as well as the Canadian Diabetes Association (CDA) numbers.[8] You'll see that the CDA numbers are more lenient, which I do not agree with. We must all be more aggressive with our diagnosis of prediabetes. The CDA reference ranges are in parentheses.

| | Diabetes | Prediabetes |
|---|---|---|
| **Fasting blood sugar** | ≥7.0% or ≥126 mg/dl (≥7.0 mmol/L) | ≥5.6%–6.9% or 100–125 mg/dl (≥6.1–6.9 mmol/L) |
| **Hemoglobin A1C** | >6.5% (≥6.5%) | 5.7%–6.5% (6.0%–6.4%) |

The side effects of diabetes are debilitating, if not deadly.

- 80 per cent of people with diabetes will die as a result of heart disease or stroke.
- Adults with diabetes are twice as likely to die prematurely, compared to people without diabetes.
- The life expectancy for people with type 1 diabetes may be shortened by as many as fifteen years; the life expectancy for people with type 2 diabetes may be shortened by five to ten years.
- Every 24 hours in the United States, because of diabetes:
    - 238 people will lose an arm or leg from amputation.
    - 120 people will enter end-stage kidney disease programs.
    - 48 people will go blind.

Diabetes is one of the most costly diseases because of the costs associated with the complications that result from it:

- The average medical costs for people diagnosed with diabetes are 2.3 times higher than for those without diabetes.
- Americans spent $176 billion on direct medical costs for diabetes in the United States. The total estimated costs of diagnosed diabetes have increased 41 per cent, to $245 billion in 2012 from $174 billion in 2007. The American economy lost $69 billion because of diabetes-related reduced productivity. This includes costs for absenteeism and unemployment caused by diabetes-related disability, and lost productivity due to early death.[9]
- By 2016, diabetes-related costs in Canada are predicted to rise to $8.1 billion. And by 2020, diabetes will cost the Canadian health care system $16.9 billion[10] every year.

## HIDDEN SUGAR AND SUGAR ADDICTION

Is the simple way of preventing diabetes to avoid sugar? It would be if sugar wasn't hidden so well in the foods we eat.

Quantity and quality of sugar play a role as well. First, our bodies are digesting more carbohydrates, or "sugars," than we've ever seen before. And not all carbohydrate calories are equal. The sugars we ingest today are not from natural sources, such as from fruit; instead, they come from processed and manufactured sugars and from sugar substitutes. Although all carbohydrates break down into sugar within our bodies, they do so at different rates and in different ways depending on the source. Our pancreas and liver can't handle these man-made sugars as readily.

Finally, ingesting sugar is addictive.

In the late 1960s, a graduate student named Anthony Sclafani noticed that lab rats quickly ate up some Kellogg's Froot Loops they accidentally discovered. He decided to design a test to measure their zeal for this sugary snack. Rats hate open spaces and prefer the corners of their cages. Sclafani put the Froot Loops in the brightly lit open center of their cages, which they normally avoid. To his surprise, the rats overcame their instinctual fears and ran out into the open to gorge on the cereal.

Years later, in 1976, Sclafani was working on another study. He was trying to fatten some rats for an experiment but was unsuccessful when feeding them Purina Dog Chow. Recalling his observations years earlier, he decided to feed them sugar-laden products—cookies and candies. The rats went crazy for them. They couldn't resist, and they became obese in just a few weeks. He published one of the first papers proving the incidence of food cravings.[11]

In the 2013 book *Salt, Sugar, Fat: How the Food Giants Hooked Us*,[12] Pulitzer Prize winner Michael Moss speaks about the food-industry term *the bliss point*. This point for sugar is the precise amount of sweetness or saltiness—no more, no less—that makes food and drink most enjoyable. At the Monell Chemical Senses Center in Philadelphia, scientists work to uncover the mechanisms of taste and smell as well as the psychology behind why we love certain foods so much. Moss interviewed scientists who conducted experiments to discover the bliss point of products for adults and children, as well as for geography and ethnic groups. The scientists discovered that children have a higher sugary bliss point than their parents. They also found that more African-Americans than other ethnic groups chose the sweetest and saltiest solutions.

The truth gets even more bitter. Most of us recognize the obvious sugars, such as table (granulated) sugar, brown sugar, confectioners' (icing) sugar, honey, and corn syrup, plus the obvious sweetness found in sugary drinks and desserts. But I'm even more concerned about the hidden sugars found in so many processed foods, such as ketchup, barbecue sauces, and "fruit" yogurts—and even peanut butter. Some estimates claim that just 1 cup (250 mL) of peanut butter contains an average of 8 teaspoons (40 mL) of sugar.

Another big source of sugar in our diets is the manufactured sweetener high-fructose corn syrup, which makes dietitians shudder. Although the traditional source of sugar coming from table sugar has dropped, we are consuming more high-fructose corn syrup, because it is a cheaper alternative to sugar and because it is hidden in foods. According to the United States Department of Agriculture, in 2011, the average American's consumption of this man-made product dramatically increased from 8 ounces (250 g) a year in 1970 to 43.5 pounds (19.7 kg) a year by 2009.[13]

High-fructose corn syrup is included in so many foods because it helps foods taste smoother, compensating for the fats, such as butter, that we've removed from our carbohydrates. High-fructose corn syrup is found in almost everything shelved on the inner aisles of the grocery store—soda, candies, jams, ketchup, and dressings. But more surprisingly, it is also found in some common "healthy" foods, such as yogurt, whole wheat breads, cereals, roasted nuts, and applesauce.

Our bodies love fructose because it is a naturally occurring form of sugar traditionally found in a piece of fruit, which also contains fiber. However, with man-made high-fructose corn syrup, we are able to obtain this sugar cheaply and in large amounts. But because it doesn't contain fiber, it is absorbed by our bodies at a faster rate. This speeds up the rate of blood sugar in the bloodstream, setting off an insulin response. The pancreas secretes more insulin to cope with the sudden sugar load. If you don't burn the sugar, you store it—as fat. And here you were, trying to avoid gaining weight as fat!

SCIENCE STAT

**Honey** If you think switching to honey in your coffee is the answer to avoiding sugar, think again. Honey is high on the glycemic index (a scale used to rate the ability of a food to increase glucose levels in the blood), so it's not a good source of carbohydrates. Although honey can be good for certain purposes (many home remedies rely on honey for some illnesses), the reality is that honey is still high in sugar.

#### WHAT MAN-MADE SUGARS DO TO US

Studies done in 2011 by Dr. Kimber Stanhope, a nutritional biologist at the University of California, show that our livers process man-made sugars differently than natural sugars. In one study, participants consumed a normal low-sugar diet for

SCIENCE STAT

**Sugar Consumption** In 2012, the USDA suggested that Americans consumed 76.7 lb (35 kg) of sugar annually. That's approximately 26 teaspoons of sugar per day or approximately 20 bags of sugar per year. According to Dr. Francesco Branco, head of nutrition for health and development for the World Health Organization, we should aim for 5 per cent of our daily calories coming from sugar if we can but 10 per cent is more realistic. This is equal to 13.3 teaspoons (65 mL) of sugar a day—based on the consumption of 2,000 calories a day.

---

a number of days; blood levels were taken to determine baseline blood levels. The next portion of the study involved swapping out 25 per cent of participants' calories for sweetened drinks containing high-fructose corn syrup. Participants' blood levels were tested every thirty minutes around the clock. At the end of the modified diet two weeks later, participants showed increased levels of LDL ("bad") cholesterol and other risk factors for cardiovascular disease.[14] What this study shows us is that when we consume too much sugar, our livers become overloaded with fructose and convert a portion of it to fat. (Recall from Chapter 4 that small and dense LDL particles are known to lodge in blood vessels, and form plaque, and are associated with heart attacks.)

If the risk of heart disease doesn't scare you straight off sugar, how about cancer? Dr. Lewis Cantley, a Harvard professor and the head of the Beth Israel Deaconess Cancer Center, claims that the hormone insulin can fuel nearly one-third of common cancers, such as breast and colon cancer.[15, 16] These tumor types have insulin receptors and encourage the tumor to find glucose to help feed its growth.

SCIENCE STAT

**Insulin** This hormone is most commonly associated with diabetes. It is secreted from your pancreas in response to how many carbohydrates you eat. Its sole goal is to keep your body's blood sugar stable—not too high (diabetes) or too low (hypoglycemia).

The classic form of diabetes is known as type 1, which is caused when the pancreas can't produce insulin, which can lead to very serious complications of significantly high blood sugar levels. The life-saving treatment is insulin injections.

What few people realize is that most people diagnosed with diabetes today have type 2 diabetes. This form of diabetes is not initially caused by the body's inability to produce insulin (type 1), but rather the body's inability to use the insulin being produced because of obesity. This is called insulin resistance. Your body produces a great deal of insulin to cope with the rising blood sugar levels, and then the pancreas tires and stops producing insulin. At that point, these diabetics require insulin injections. As their insulin resistance develops, a lot of damage is done inside their bodies. The high blood sugar levels and high insulin levels affect the other hormones. In particular, growth hormone levels drop.

What can be done as society battles an epidemic of newly diagnosed diabetics in the developed world? The combination of poor food choices, lack of physical activity, and our hormonal response to stress are all accelerating this largely preventable disease.

A few years ago, I wrote an article for *Canadian Business* titled "Skipping meals makes you dumb and fat." It's true. When you wake up in the morning, cortisol is secreted to get your blood pressure up and to get your heart rate going. But if you walk out of your house without breakfast, two things happen. Without food fuel, your body begins to secrete more cortisol as a stress response to being starved and your blood sugar levels decrease, slowing down your brain function. Next, this cortisol surge causes a rise in the secretion of insulin. This hormone gives you a sugar craving to find carbohydrates—fast food fuel. (You'll never find yourself feeling stressed and craving a hunk of meat, but certainly a chunk of chocolate or candies.) A high intake of carbohydrates can lead to obesity because insulin stores the excess sugars as fat.

Glycation is caused by poor sugar control. In effect, abnormal levels of insulin indicate that your body is overwhelmed by the sugar load from carbohydrates in the diet. As well, the fat that accumulates in our bodies, especially around our internal organs (known as visceral fat), contributes to damaging inflammation too. The only way to lower the glycation and inflammation caused by excessive insulin is to cut down on your intake of carbohydrates and to burn off excess fat with exercise.

---

### HOW ABOUT SUGAR SUBSTITUTES?

In July 2013, Susan Swithers, a behavioral neuroscientist at Purdue University, took aim at diet beverages and the reputation their manufacturers promote as healthy drink alternatives. In her mission, she scrutinized the most recent research on high-intensity sweeteners. She wanted to show that, despite the sweeteners' low-calorie contents, these drinks could cause a host of other health problems in those who consume them, such as weight gain and overeating. She focused on drinks that contain aspartame, sucralose, and saccharin (approximately 30 per cent of American adults regularly consume these sweeteners in their diets). Following these evaluations, she suggested that diet soda may actually be as unhealthy as non-diet soda.[17]

Previous studies pointed to the same conclusions. For example, in the 1990s, the San Antonio Heart Study reported that artificially sweetened beverages increased body weight in adults and teenagers compared to the same demographics who consumed beverages that had been sweetened with traditional sugar.[18] And a host of other studies, including the Nurses' Health Study and the Health Professionals Follow-Up

Study, reported greater risks for type 2 diabetes, heart disease, high blood pressure, and metabolic syndrome (a combination of disorders that, when they occur together, increase the risk of diabetes and cardiovascular disease) in people who consumed artificially sweetened beverages.[19]

There are a few ideas out there that try to explain why this is happening. Researchers have explored the issue in animal studies and found that there's a link between overeating and the consumption of artificial sweeteners, such as aspartame and saccharin.[20]

Is it possible that artificial sweeteners affect the body's ability to track calorie intake and needs? In other words, perhaps when the mouth tastes something sweet, the digestive system expects to receive calories and fat, but these never materialize, confusing the body's metabolism and leading to further caloric intake from other foods.

On a psychological level, it's not difficult to imagine that people feel they can splurge in one area if they have saved in another. This cognitive distortion, as some researchers call it, translates into consumers indulging in fat-heavy foods because they feel they are entitled to enjoy it—having just consumed fewer calories in their beverage. Some people call this the "Diet Coke and French fries phenomenon."

### GOOD FAT, BAD FAT, LOW FAT, FISH FAT

We've made the sugar situation worse with manufactured high-fructose corn syrup. What's more, we've done a similar disservice to fat. As a result of one simple dietary recommendation to consume only low-fat or non-fat foods, we've caused another problem: the dramatic decrease in essential, or "good," fats in our diet. (Recall that essential fats are strong anti-inflammatories.)

Many good fats can be found in fish. Rarely do I hear of a family eating fish regularly. All too often, parents eat very

**Sources of Omega-3 Foods**

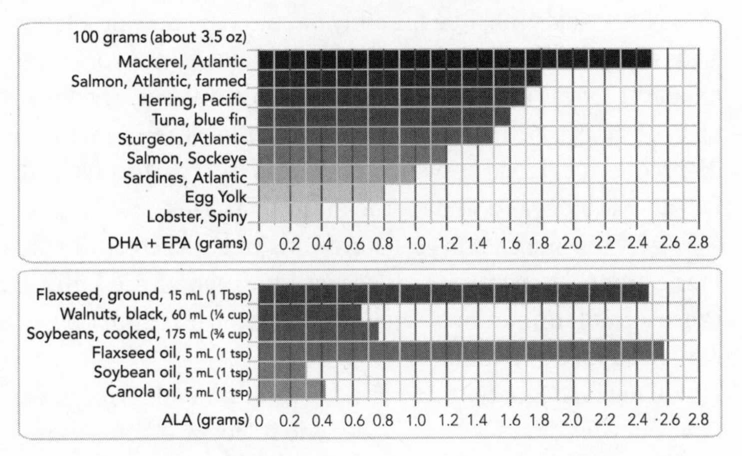


There are three types of omega fatty acids: DHA, EPA, and ALA. All are important for maintaining good health. DHA and EPA are generally found in fish, while ALA is found in nuts and seeds (and their oils).

little, and children eat none. Fish fats contain PUFAs (polyunsaturated fatty acids), or omega-3 and omega-6 fats, which break down into DHA (docosahexaenoic acid) and EPA (eicosapentaenoic acid). These critical nutrients are often found in very low amounts when I measure them in my patients. It's only the ones who eat fatty fish (cold water fish, such as salmon, sardines, and rainbow trout) at least three times a week who seem to have enough DHA and EPA in their bloodstream. These acids can be found in plants as well. But they don't have the same properties as the fatty acids that come from fish. Other high-PUFA foods include nuts, avocados, and certain oils (sunflower, grapeseed, and olive). Unfortunately, they seem to drop out of our diets as we eat more processed meals. I have more to say about these nutrients in Chapter 6.

One benefit of omega-6 fatty acids is that they can prolong the shelf life of processed foods. And we need these fatty acids: they are integral to our nutritional health because they play a role in controlling how our cell membranes work. But these fatty acids must be properly balanced by omega-3 fatty acids to work well. And the bad news is that we don't consume enough omega-3s to achieve this balance. In fact, the ideal ratio of omega-6 to omega-3 is 3:1—yet the North American diet scores a balance between 15:1 and as high as 30:1. We eat far too few omega-3s and far too many omega-6s.

When too many omega-6s are consumed, the cell membrane produces chemicals called cytokines, which are a direct cause of inflammation. You'll recall that inflammation is really bad for our telomeres.

We need adequate levels of DHA in our blood. Why? DHA deficiency is connected with a decline in how well our brains function, and it's been linked to degenerative conditions such as Alzheimer's disease. In other words, this could be a big concern as we age. One mechanism connected to this decline may be phosphatidyl serine (PS). This is a phospholipid component that controls the auto cell death of weakening cells. What does this mean? Low DHA levels lower our neural cell PS and quicken our neural cell death. As well, low levels of DHA are found in severely depressed people. Could low PUFAs be causing the increase in mental health problems in our society today? Perhaps.

SCIENCE STAT

**Association between Fish Consumption and Depression** A study found that, in women, each additional weekly serving of fish eaten decreased the risk of having a new bout of depression by 6 per cent. And those who ate fish two or more times per week had a 25 per cent lower risk of depression than those who ate less per week.[21]

The richest sources of essential fatty acids are oily fish, such as mackerel, herring, salmon, trout, and sardines. So strict vegetarians and vegans are more likely than meat eaters to have low levels of DHA. This is because foods such as nuts, seeds, dark leafy greens, and whole grains contain as little as 1 per cent of the DHA that oily fish contain, and the only DHA-rich vegetable source is algae. (It's important not to consume too many large fish, such as red tuna, shark, or swordfish, because they contain high levels of mercury.)

---

**The Skinny on Fat, True or False**

TRUE OR FALSE: All fats are bad.
*False:* Monounsaturated fats and polyunsaturated fats lower your cholesterol and can cut your risk of heart disease, so these fats are good for you. Saturated fats and trans fats are the bad fats; they raise your cholesterol level and your risk for heart disease.

TRUE OR FALSE: Cutting fat from your diet is crucial.
*False:* You need some fats—just make sure you're eating more of the good fats than bad fats. The right mix is important for your health.

TRUE OR FALSE: Eating fat-free foods might harm you.
*True:* Many fat-free foods are packed with sugar, refined carbohydrates, and calories, so just because a package says it's fat-free does not mean you can eat all you want. Tread carefully: the term *fat-free* can be a dieter's nightmare. Low-fat yogurt is a good example; you could forfeit fat and gain sugar. Eating Greek yogurt may be a smarter choice. It contains roughly half the carbs of the regular kind of yogurt (5–8 g per serving compared to 13–15 g).

TRUE OR FALSE: It pays to eat a low-fat diet.
*False:* Sadly, it seems the opposite is true. Obesity rates have doubled since the 1980s, which was the same time we saw the first low-fat foods lined up on grocery shelves. Eating the right fats makes you feel full, which can stop the impulse to overindulge.

## Nutrient-Deficient Diets

I wrote earlier about how a lack of DHA can be bad for your brain health. But sometimes when we try to do something good for our health, we unknowingly deprive ourselves of the nutrients our bodies really need.

A developing trend in the low-fat diet craze, for example, is the idea of becoming vegan (a form of vegetarianism that eliminates all forms of animal products from the diet). Although vegetarianism is common in many parts of the world, the key to following this diet correctly is to avoid potential pitfalls. Lowering fat intake, losing weight, or avoiding animal products for religious or ethical reasons can translate into a drop in protein intake. This results in nutrient and hormone deficiencies. But we all need protein because our bodies convert it into three essential substances: enzymes, new proteins to be stored as muscle, and hormones—and none of us can function without these three things.

So what are the important components of proteins? Amino acids—the building blocks—are classified as essential amino acids (required for our bodies from our diet) and non-essential amino acids (ones that our bodies can manufacture from other amino acids). It's possible to obtain all eight essential amino acids from traditional types of animal proteins, but vegetarians will find this task more challenging because those amino acids aren't as easily tapped into by eating vegetables and legumes. Quinoa is on the rise as a popular grain for this very reason: it is the only one that contains all eight essential amino acids.

Legumes, such as peas, peanuts, and beans, are alternative sources of nonanimal protein. White meat poultry is also an excellent alternative (red meats generally cause more inflammation). In fact, we need to focus on *all* sources of protein, making sure we have enough fish (or an omega-3 equivalent) in our diet. Why?

If we don't ingest enough protein, our bodies become low in essential amino acids, which are critical for producing hormones and enzymes. In my patients, in addition to amino acid deficiencies, I commonly see deficiencies in vitamin $B_{12}$, folate, and iron among "recreational" vegetarians (those who choose to dabble in this way of eating without informing themselves about how to be a healthy vegetarian).

FURTHER INFORMATION

**What Is a Balanced Diet?** Drs. Edmund Chein and Hiroshi Demura point out in their book *Bio-Identical Hormones and Telomerase* that the traditional Japanese breakfast—miso soup, eggs, and steamed rice—is a great example of a balanced meal.[22] Between the three elements, all eight essential amino acids are covered. This supports the formation of good hormones—not to mention the healthy protein from the eggs. The extraordinary lifespan of the Japanese is, in part, due to their diets.

### What Do Your Teeth Want to Chomp On?

Do you ever think that we were genetically designed to eat meat? The clue is in our teeth. Our eight premolars and twelve molars (including wisdom teeth) work to grind grains, seeds, vegetables, and fruits. Our eight incisors are there to help us bite and tear into foods. Note that we only have four canine teeth to cut meat. For this reason, I believe our mouths are sending us a message: we do need meat, just not that often.

Moderation is key. After all, when humans had to hunt for meat, they weren't always successful. It's not like the meat came right to them as it does for us today in the grocery aisle. People who live the longest (such as those in Japan, Sardinia, and Costa Rica) share a common trait: they don't eat anywhere near as much meat as the average North American does. When they do serve meat, it's usually on special occasions.

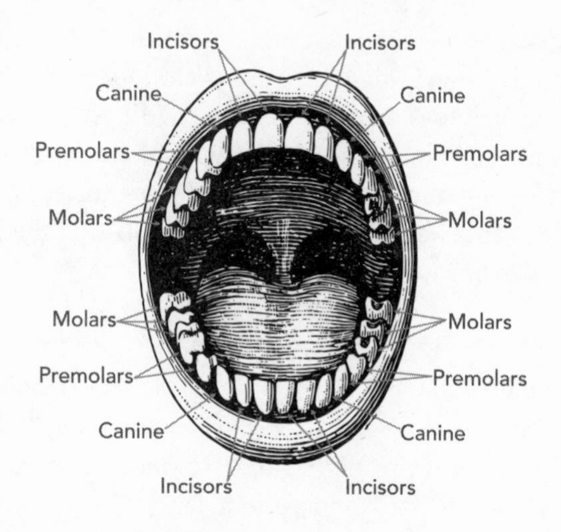


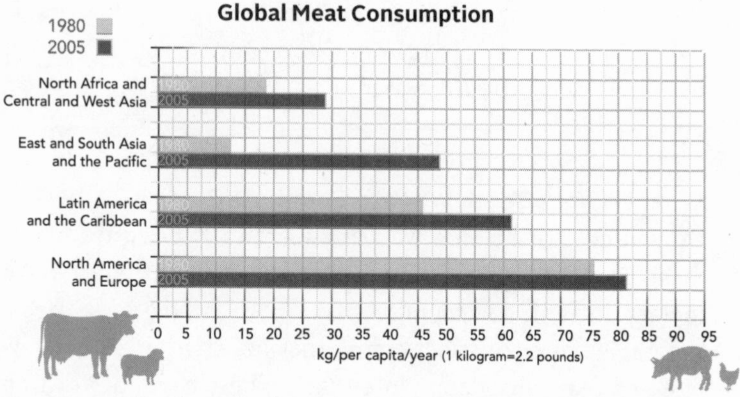


The consumption of meat has been rising all over the world, most notably in the developed countries.

## The Salt of the Earth

The phrase *salt of the earth* comes from the Bible, Matthew 5:13. "Ye are the salt of the earth: but if the salt have lost his savour, wherewith shall it be salted? It is thenceforth good for nothing, but to be cast out, and to be trodden under foot of men."

**Percentage of Salt Found in Certain Types of Foods**

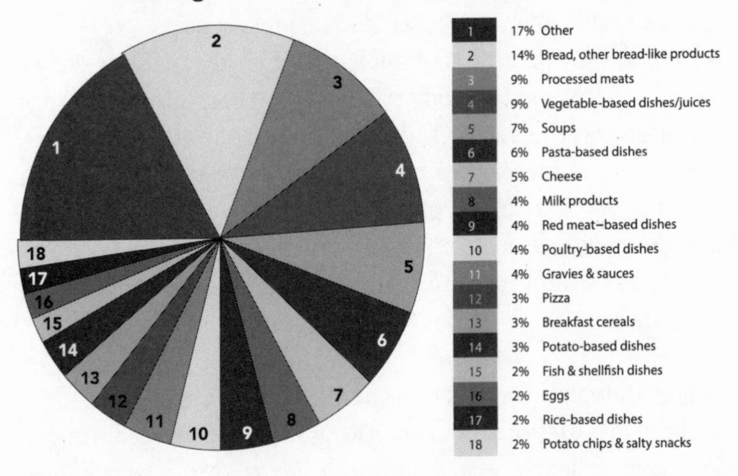


This idiom is used to describe people of great worth and reliability. It seems that *salt of the earth* was a phrase coined in reference to the value of salt at the time.

These days, salt has a bad rap. But salt, in itself, isn't a bad thing; it's the amount of salt we consume that is detrimental to our health. Our body can't produce salt on its own. Yet it requires this crystalline compound for a number of essential functions: to control our blood pressure, to transmit information between our cells in nerves and muscles, and to absorb nutrients from the small intestine during digestion.

Most of us don't realize that we're consuming too much salt. North Americans ingest about 3,400 mg of salt daily. But health guidelines recommend we take in less than 2,300 mg. This is equal to a single teaspoon of salt.

According to the Federal Drug Administration, 75 per cent of the dietary sodium we consume comes from packaged and restaurant foods, not from the salt shaker. Some of the biggest culprits include cheese, processed meats (especially bacon), tomato sauce, salted cod and mackerel, anchovies, baked and refried beans, canned legumes, breads and rolls, and, of course, the salty condiments (like soy and teriyaki sauces).

When we consume too much salt, it throws off our body's fluid balance, and the body tries to bring back proper balance by secreting the excess sodium. This essentially dehydrates the body and increases a hormone called angiotensin, which tends to increase blood pressure. And high blood pressure is a major risk factor for stroke, heart disease, and kidney disease. So it's important to read the food labels to count your salt intake.

### Buyer Beware: Today's Hot Diet Is Tomorrow's Regret

Weight-loss diets are not something I would ever recommend. We have to learn as a society to change our eating habits. Period. This is a behavioral change that all of us need to adopt. One study, published in *American Psychologist* in 2007, merged the results of thirty-one earlier dieting studies and came to the frightening conclusion that, if anything, dieting is often a predictor of future weight *gain*.[23] In fact, the study showed that up to two-thirds of dieters end up gaining more weight than they lost in the first place.

DEFINITION

**Caloric Restriction** Caloric restriction is not the same thing as dieting or starving yourself. This is evident in the shorter lifespans of people who live in developing countries where starvation is a day-to-day reality. Rather, healthy caloric restriction refers to ingesting foods that have the highest nutritional value in the fewest calories possible, based on your activity levels.

We read about the benefits of maintaining nutritious, healthy eating habits—and about the health risks of doing the opposite. And research has also shown us that restricting calories can extend lifespan—at least with mice. In these animal studies, mice lifespans increased from 41 months to 56 months (120 to 150 human years) by following calorie-restricted diets. Scientists believe that calorie restriction may increase our lifespan by changing the levels of melatonin and growth hormone. Read more about melatonin and growth hormone in the coming chapters.

### BUT DOES IT WORK IN HUMANS?

To determine if the benefits of a restricted-calorie diet could also extend the lifespan of humans, scientists conducted a fascinating experiment in the early 1990s at a science research facility called Biosphere 2 in Arizona. For two years, scientists lived inside an air-locked structure where they cultivated and raised their own food sources. The foods they consumed were nutrient-rich, but they consumed 30 per cent fewer calories than the average moderately active American male—about 2,500 calories per day.[24]

At the end of the two-year period, doctors evaluated the health of the Biospherians and found that they had lower cholesterol, blood pressure, and fasting blood sugar levels than when they entered the facility. Even after the scientists returned to their "normal" lives and went back to consuming foods as they did prior to the study, these levels remained low. It will be interesting to see what effects this study will have on the health of these participants in the long term. But at least for now, it appears that the benefits of a restricted-calorie diet do help us live healthier lives.

## The Beauty of Water

Do you take a shower or bath every day? Or use moisturizers on your face? We wash our bodies with water and we put lotion on our skin, yet many of us forget the importance of water for the inside of our bodies. Every body is different, and so are every body's water requirements.

The amount of exercise and activities you do, where you live, and your sex and age all affect your hydration needs. But despite the varying requirements of different people, everyone needs water.

Water makes up 60 per cent of our body weight. And because our bodies are constantly losing water (through aspiration, perspiration, and urine and bowel movements, among other functions), we need to replenish that water by drinking more of it.

In fact, water is important to every one of our body's essential functions. It carries nutrients to your body's cells; aids in digestion; acts as a lubricant for mouth, nose, and ear tissues; improves skin tone and healthy muscles; flushes toxins from the body; helps the body fight bacteria and viruses that lead to inflammation and oxidative stress; and much more. And I've already touched on how inflammation and oxidative stress affect our telomeres.

Different health organizations recommend varying amounts of water consumption. One of the most reliable, the Institute of Medicine, recommends that men drink 3 liters (13 cups) of total beverages every day, and women drink 2.2 liters (9 cups). Beverages do not include caffeinated and alcoholic drinks because they dehydrate you. In fact, you should drink additional water when you consume such beverages. Food can also contribute to your hydration levels. According to the Mayo Clinic, food provides approximately 20 per cent of a person's total water intake.

If you're not sure if you're drinking enough water, pay attention to the color of your urine. The more yellow it is, the more dehydrated you are. Urine should be colorless or light yellow. Sip water throughout the day to stay hydrated—rather than drinking the recommended amount all at once.

There are other factors that can influence the color of

your urine, including some foods (beets and blackberries, for example), certain medications, and some vitamins (such as the B vitamin spectrum). Speak with your health professional if your urine is too dark.

### To Eat Better, Look at What You're Eating Now

Up until now, this chapter has been a survey of the greatest concerns facing you as you try to meet one of your most basic needs. So, now the question is, what do you have to do to ensure that you are eating in a way that will keep you in Peak Health? To guide you in eating more healthfully, you need to assess your current diet. Start with a nutrient profile by keeping a chart of your diet for one week. Many websites offer programs that help you count calories and, more important, understand the relative proportions of carbohydrates, proteins, and fats.

FURTHER INFORMATION

My recommended nutrition websites include:
myfitnesspal.com
sparkpeople.com
nutritiondata.self.com

---

I can't emphasize enough that each of us is different and that you need to understand what makes your body tick before you can decide what's best for you to eat. When you get your diet right, your telomeres benefit. True, you can't see your telomeres the same way you can see an expanding waistline, but if you're overweight or obese, you know that you're exposing your body to glycation, inflammation, and oxidation. As a result, you can be sure your telomeres are ground down shorter with every day that passes.

How can you find out what your body really needs?

What the proportion of carbohydrates, proteins, and fats looks like for you will depend on your health status and your health goals.

### NUTRIGENOMIC TESTING

In order to fully sort out the personal diet a person needs, I often suggest nutrigenomic testing for my patients in order to open their pathway to personalized nutrition. There are many identified genetic markers that can tell us how our bodies process sugars, fats, nutrients, and vitamins.

In my practice, some patients have enough exposure time to the sun to produce vitamin D directly from their skin. Yet I see from their blood measurements that their vitamin D levels are low. Why? Genetically, they have a mutation of their vitamin D receptor. In other patients, I find a low level of omega fatty acids. Although they eat fish four or five times a week (which should be more than sufficient), their genetics reveal a difficulty in absorbing omega fatty acids from food sources.

You can also measure all the nutrients in your body and understand your functional needs for them.

Then you'll want to undergo food allergy and sensitivity testing. You certainly don't need to eat foods you know your body is "rejecting" biologically. Eating these foods only increases inflammation, which you'll recall is not good for your telomere health.

Finally, you'll want to get some professional help from a dietitian or a qualified health care professional who has been trained in nutrition to create a diet personalized for you.

## Food Fueling

Your body converts carbohydrates, proteins, and fats into energy at different rates. Do you ever wonder why nutrition-

**Food Energy Sources**

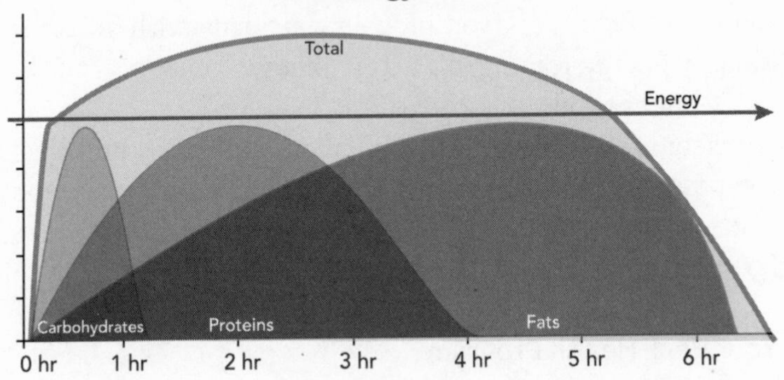


ists always recommend that we eat a balanced meal? What does this really mean?

There are three types of food fuels.

1. *Carbohydrates:* These plant-based foods provide an almost-instantaneous source of energy in the form of sugars. The energy comes quickly, but it also fades quickly. Examples include grains (such as wheat, rice, oats), fruits, and vegetables.
2. *Proteins:* Proteins come from both plant and animal sources. They provide the same amount of caloric energy as carbohydrates, but are not as readily available to your body. They provide an intermediate source of energy. Examples of plant sources are chickpeas and quinoa. Animal sources include meats (such as chicken, beef) and seafood (such as fish, shrimp).
3. *Fats:* Also from plant and animal sources, fats provide higher caloric energy than both carbohydrates and proteins. However, this food fuel supplies your body with energy hours after you ingest and digest it, and for a longer time. Examples of plant sources are cooking oils (such as olive, grapeseed, avocado). Animal sources include dairy (such as milk, cheese, butter) and fat attached to meat.

When you eat a balanced meal that contains all three sources of food fuels, you provide your body with all three kinds of energy, leading to a better energy picture over a five- to six-hour period.

There is something to be said about a farmer's breakfast composed of eggs, homemade sausage or peameal bacon, and whole-grain toast! But it's only helpful if you are active on a farm and not sitting at your desk.

### Your Peak Health Program

Your diet is your most important arsenal in achieving your Peak Health. Follow these steps, and you will be well on your way:

1. Log your food intake.
2. Do your nutrigenomics by way of a blood test or a buccal swab.
3. Get your nutrient status profile done by way of a blood or urine test.
4. Have your food allergies and intolerances tested by way of a blood test.
5. Most importantly, get a health professional who is familiar with all of these components to put them all together for you to create your customized eating plan.

# 6

# Supplements to Counteract Your Deficiencies

## Why Do We Need Supplements?

It's almost impossible for anyone to get enough of the right nutrients from food these days. And even if we consume enough nutrients and vitamins in our diet, many of us can't absorb the nutrients we need because of stress, allergic reactions, inactivity, or medications. What's more, as we get older, our bodies become less efficient at digesting and absorbing. Let's consider whether we should use supplements, and if so, how we can supplement our diets in order to achieve Peak Health.

But isn't our food much more nutritious than years ago?

Emphatically, no. Our sources of healthy foods aren't as nutrient-rich as they once were because the soil that good foods are grown in is often depleted of minerals. And what are the effects of pesticides and fertilizers on our food chain? Terrible. Even worse, our foods are then shipped to factories, packaged, and processed into completely new "foods," which, in essence, depletes the foods of what makes them healthy. In fact, some experts claim that even the most healthful diet—from any culture or geographic area—could never contain enough antioxidants to truly protect the body from the damage done by free radicals and oxidative stress.

Our bodies require vitamins and minerals to survive, and most of these elements can't be produced by our bodies alone. Many of those same experts believe that we can only reach optimal health by complementing a healthy diet with supplements. On the other hand, a much smaller group of experts claims that using supplements can lead to an increased risk of death.

But can your good health come in a bottle? It's estimated that people spend more than $60 billion every year worldwide on vitamins, supplements, and herbs. Americans spend about half the global total—around $30 billion. Canadian demand is growing as well; Canadians spent more than $3 billion on supplements in 2012. That's a lot of money and belief in something found in bottles. One thing's for sure: supplements deserve the discussion and attention they are getting.

## Can You Count on Supplements to Live Healthier and Longer?

There are three camps of believers when it comes to the use of supplementation. Naysayers believe that you will get enough nutrition if you just eat the right meals and that taking supplements is a waste of money. Well, I've already explained why that argument doesn't work anymore. The pro-supplement users will point to studies that suggest that taking some form of vitamins and minerals will help with our health and longevity. Then there are those in the middle ground who select a few supplements here and there, just in case.

SCIENCE STAT

Most of us don't eat well enough. In fact, one study reviewing the eating habits of Americans found that only 1 per cent of adults met all five food group recommendations[1]—namely grains, fruits, vegetables, proteins, and dairy.

Believe me, Canadians eat no better. According to a 2012 Health Canada survey[2]:

**Health Canada Survey 2012**

5 in 10 women, and 7 in 10 men, have energy intakes (food) that exceed their energy needs (activity).

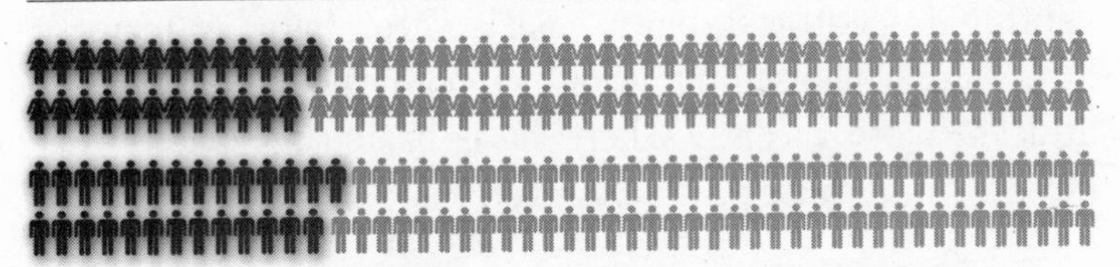

25% of males and 23% of females who are 19 years and older have fat intakes above the Acceptable Macronutrient Distribution Range (AMDR).

32% of males and 21% of females who are 19 years and older have carbohydrate intakes above the AMDR.

Here's my view: supplements do have benefits. But you can't ask one or two supplements to perform miracles and be the magic bullets to prevent or treat your heart disease, diabetes, or cancer. This is asking too much of any supplement or herb.

And, as with anything on the topic of our health, it's complicated. Some studies that initially showed promise have fallen short. Here are a few examples:

- Vitamin E: It's been touted to prevent heart disease, but a large Heart Outcomes Prevention Evaluation (HOPE) study showed the opposite may be true. It also found that vitamin E may have promoted an increased risk of heart disease in the study participants.[3]
- Vitamin D: This vitamin was thought to help prevent bone fractures, but one study showed no prevention of fractures in older women; in fact, it actually showed a higher rate of fractures in the women studied.[4]
- Omega-3 (fish oil): This supplement shows promise in preventing heart attacks, but it has also been found to increase LDL ("bad") cholesterol levels[5]—a potential risk for heart attacks. More recently, there have been suggestions that omega-3 may cause prostate cancer.[6]

These studies may discourage you from using supplements altogether. Please, don't be discouraged. Without a full arsenal of essential nutrients in your body at all times, your body won't work properly. If you leave your body in a chronic state of nutrient deficiency, your health will be compromised. For this reason, it pays to consider the benefits of supplements as a way to augment your diet. My advice? Take supplements only if you know you are deficient in a vitamin or mineral, in an organic amino acid, or in fatty acids. All these nutrients are measurable via blood or urine tests. This is the best way to identify your deficiencies—and your needs. But one caution: you shouldn't take supplements without the guidance of a professional, such as a naturopathic or osteopathic doctor, who knows how to use them properly.

In my medical practice, I have found nutrient deficiencies in hundreds of my patients through simple lab tests. Many of these nutrients are critical in preventing glycation, inflammation, and oxidation, as well as in the regulation of hormones. All of these can affect the health of your telomeres.

Before these patients had these tests, many believed they were eating properly. As I've pointed out, eating healthily can be hard to do. Your food habits and routines, preferences, intolerances, allergies, and lifestyle, as well as the season, can all affect what food you can access. Deficiencies most often occur in wintertime, or with vegetarians, and, yes, with picky eaters as well. And poor soil nutrients lead to nutrient-deficient foods, which compounds the problem.

These realities make a pretty good case for supplementation to at least augment what we absorb from our diets. And this brings us back to my initial question: to supplement or not to supplement?

And what if your goal is not just to prevent or cure disease, but also to promote healthier cells and telomeres?

## Omega Fatty Acids and Telomere Length

There's solid research about nutrients and telomeres. Let's start with omega fatty acids. You will recall from the previous chapter that cold water fish is high in this anti-inflammatory nutrient. A study conducted by Drs. Elizabeth Blackburn and Mary Whooley examined a group of aging patients with coronary artery disease over a five-year period. They discovered that people who consumed the least amount of omega-3 fatty acids had the fastest rate of telomere shortening.[7]

## Multivitamins and Telomere Length

Here's another study that deserves attention. Scientists from the National Institutes of Health showed that longer telomere length in women is associated with consuming multivitamins.[8] Researchers from the Sister Study evaluated data from 586 subjects aged 35 to 74 years. After ruling out other factors, such as age and other potential factors, the group found that the telomeres of multivitamin users were more than 5 per cent longer than nonusers'. The study also determined that those who regularly consumed vitamins E and C had longer telomeres than those who did not (even after adjustments were made to reflect the use of multivitamins).

### Vitamin D and Telomere Length

Our telomeres love sunshine too. Studies show that telomere length is greatly affected by vitamin D, which we absorb from certain foods in our diet, such as fish, oysters, eggs, mushrooms, and fortified cereals and dairy products, and from the sunshine on our skin. In fact, the telomere length between the lowest and highest vitamin D groups translated into five years of telomeric aging—even after adjustments were made for age, seasonality, physical activity, hormones, and other factors.[9]

SCIENCE STAT

**Bioavailability** Different substances are broken down at different rates in our bodies. *Bioavailability* refers to the degree and rate at which a substance (such as a drug) is absorbed into the body. There are a few advantages to measuring bioavailability in white blood cells instead of through our blood and urine. For example, testing white blood cells:

- Provides a more accurate measurement of long-term *functional* nutritional levels (over four to six months).
- Can indicate if nutrients are being absorbed and functioning properly. This, in turn, reflects your total metabolic function.

## Four Supplement Tips for Your Telomeres

We don't know yet if supplements can signal our telomeres to "power on" their replenishing telomerase to protect our telomere length. But we do know that maintaining ideal nutrient levels is crucial to our good health. So it stands to reason that the right nutrient levels will benefit our telomeres as well. Here are four quick tips for getting supplementation more right than wrong.

1. Get tested. All nutrients—vitamins, minerals, fatty acids, and organic and amino acids—can be measured in your blood and urine. They can also be tested by measuring the bioavailability of micronutrients in your white blood cells, which is the method I use in my clinic.
2. Eat a healthy diet. If you know you don't eat well, at least take a multivitamin, which will provide some protection against deficiencies. Just know it will never replace a healthy food lifestyle.
3. Select a daily multivitamin. Although quality multivitamins aren't cheap, in the big picture, they remain an inexpensive preventive health measure. Take one at the same time every day so that it becomes a habit. Make sure to pick a reputable "professional" brand, because the company will guarantee excellent manufacturing practices and source raw materials from safe suppliers. Most important, these supplements will be more bioavailable, will be toxin-free, and won't make false claims about the contents. The Mayo Clinic recommends choosing a multivitamin that provides the recommended dietary allowances of vitamins and minerals (vitamin C, vitamin E, B-spectrum vitamins, calcium, chromium, copper, iron, magnesium, selenium, and zinc). But what should determine what your supplemental needs might be is your current state of health. So you should discuss it with a doctor who is well educated on supplement use.

4. Avoid the "megas." It's also important to steer clear of the mega-dose vitamins and mega-fortified food products, unless directed by your health professional. Too much of a good thing can be trouble too. And remember, if a supplement brand makes unrealistic promises in their marketing materials, it's probably safe to assume their claims are false. Don't fall for the "super" supplements.

A nutrient profile obtained from blood and urine samples is a powerful tool in enabling a health professional to understand the precise needs of their patients. This takes the guesswork out of wondering if you have enough nutrients that are critical for proper telomere function. There are so many supplements to choose from—an incredible 54,000 are listed in the Natural Medicines Comprehensive Database. So you must test yourself first to know which brand to buy.

I asked my colleague Dr. Shelley Burns, a naturopathic doctor, to give me her top ten supplement picks and to explain why she swears by them.

**Top Ten Supplements as Selected by Dr. Shelley Burns**

When it comes to Peak Health and peak performance,. it's important to consider the nutrients you need to prevent glycation, inflammation, and oxidation, and to help promote healthy hormone function.

There's definitely an overlap in the function of different nutrients. For example, alpha lipoic acid and the B vitamins reduce glycation, but the B vitamins are also associated with reducing inflammation along with the omega-3 fatty acids, vitamin C, and vitamin D.

As for oxidation, alpha lipoic acid, coenzyme Q10, and vitamin C are all powerful antioxidants. Iron can also be considered an antioxidant when used in small doses; however, in large doses, it can become an oxidant. Magnesium can prevent both glycation and inflammation. The amino acids glycine, glutamine, and arginine all prevent oxidation and glycation. As Dr. Chin says, you need a professional to help you determine what your unique needs are. Having said that, here are the top ten supplements I recommend.

### 1. OMEGA-3 FATTY ACIDS, EICOSAPENTAENOIC ACID (EPA), AND DOCOSAHEXAENOIC ACID (DHA)

Omega-3 fatty acids are beneficial in reducing inflammation in your body. There is strong evidence the omega-3 fatty acids can improve your heart health, decrease joint inflammation, reduce the intensity and number of headaches, and reduce depressive symptoms. The only downside to omega-3 fatty acids is that they are also considered a natural blood thinner, so bruising may occur if they are taken in high doses.

What becomes confusing is the amount of omega-3 fatty acids you need to help counteract the inflammatory process. We know the omega-3 fatty acids that come from plant sources convert into ALA (alpha linolenic acid) and provide only a small amount of EPA. This is the potent anti-inflammatory component of fatty acids and therefore does not provide anti-inflammatory benefits.

Fish oil has the highest concentration of EPA. However, not all fish oils are created equally. The common daily dose of a fish oil capsule is 1,000 mg, but this does not mean you will absorb 1,000 mg of EPA. Some capsules will only liberate 250 mg of EPA and others will liberate as much as 600 mg. A therapeutic dose is between 2,000 and 4,000 mg of EPA.

Let's do the math: you require eight capsules containing 250 mg EPA, or five or six capsules containing 375 mg EPA—every day.

When purchasing an omega-3 fish oil, go for the oils that are blended—salmon, sardines, anchovies, and mackerel. Be wary of just salmon oil (due to the high mercury content).

Also, pure cod liver oil will have a higher concentration of vitamin A, which can increase the risk of lung cancer in smokers.

### 2. COENZYME Q10

Air pollutants, toxins, cigarette smoke, cell metabolism, exposure to the sun, and other environmental factors initiate free radicals, which can cause dangerous reactions that destroy our cells and damage DNA, proteins, and fats.

Coenzyme Q10, also called CoQ10, ubiquinone, or ubiquinol, is a fat-soluble nutrient present in virtually all cells and is considered the "spark plug" of the body. CoQ10 has three main functions: to act as a potent antioxidant (the antidote to free radicals), a membrane stabilizer, and a cofactor in building many metabolic pathways, particularly in producing energy. CoQ10 also helps regenerate other antioxidants in the cell, especially vitamin E.

Your CoQ10 levels are highest during the first twenty years of your life and decline with time—so much so that at age 80, your CoQ10 levels may be lower than they were at birth. Yet your body's demand for CoQ10 increases with age. Furthermore, statin (cholesterol-lowering) medications can further deplete your body of CoQ10.

The recommended daily CoQ10 dose is 30 mg, in combination with alpha lipoic acid and vitamins A, C, E, and selenium. Foods highest in CoQ10 include organ meats (e.g. liver), sardines, beef, peanuts, spinach, and albacore tuna. However, it would take 1 lb (500 g) of sardines every day, 2 lb (1 kg) of beef, or 2.5 lb (1.25 kg) of peanuts to provide 30 mg. It's important to know, too, that cooking foods at high temperatures degrades the enzyme. So you'll likely need to supplement your CoQ10 to achieve any real therapeutic effect. CoQ10 can also benefit the body if it's applied topically on your skin; it's a small molecule that can easily penetrate the skin.

When CoQ10 is combined with vitamins C and E in creams or lotions, the synergistic effect can neutralize free radicals, thus reducing wrinkles.

You should consider purchasing CoQ10 supplements in the form of ubiquinol instead of ubiquinone. Ubiquinol has a greater bioavailability than the ubiquinone form, and is the form that has the ability to scavenge free radicals. Ubiquinol, however, costs more and doesn't have as long a shelf life as ubiquinone.

### 3. VITAMIN D

Weather plays a role in our ability to get enough sunlight to manufacture vitamin D from our skin cells. Living in northern climates (such as Canada and the northern United States) and dealing with the elements of winter can impact our nutritional status—and one vitamin in particular: vitamin D, our sunshine vitamin. There is so much research on its benefits that it's hard to discount its importance to our biochemistry. We know the sunshine vitamin is essential for directing calcium to our bones, but that's not where the story ends.

Vitamin D has also been linked to lowering the risk of cancer, including colon and breast cancers. It can improve insulin sensitivity, further decreasing the risk of diabetes (and glycation risk), and it may also strengthen our immune systems and help improve our mood during the winter months.

Another interesting consideration about vitamin D is how it acts more like a hormone than a vitamin. How so? For the most part, your body can't make vitamins on its own. However, it can produce its own vitamin D through ultra violet (UV) rays—sunshine. That said, some people have a genetic predisposition that stops them from generating their own vitamin D from the sun. If the vitamin D receptor gene (VDR) is "turned off" in a person's skin, they can't make vitamin D. In this case, supplementation is a must. Getting the right dose of vitamin D depends on many factors—age, gender, race, and the environment—and the "right dose" can range from 2,000 IU (international units) daily to upwards of 10,000 IU daily. It's important to ingest vitamin D supplements with food; vitamin D is a fat-soluble vitamin and absorbs better with a meal.

### 4. MAGNESIUM

Magnesium is an important trace mineral that's essential to over three hundred biochemical reactions in your body. Some of these reactions include energy production, bone building, the synthesis of fats and antioxidants, wound healing, and the secretion of certain hormones. A magnesium deficiency can lead to a number of health problems, such as muscle cramping/twitching, headaches, irritability, constipation, and decreased bone strength.

But magnesium has other benefits as well. It can be used to decrease blood pressure or as a sleep aid. And it can also reduce the number of asthma attacks, and cut the risk of metabolic syndrome associated with an increased risk of diabetes.

You can develop a magnesium deficiency when you don't consume enough via your diet. But it can also occur if you exercise strenuously. When we exercise, we sweat (an important process in keeping our bodies cool—our own internal air conditioner). The downside to our "air conditioning unit" is that it depletes our magnesium supply and triggers the symptoms I've just described.

It's important to replenish with an electrolyte replacement or magnesium tablets during exercise and after exercise. Another way we deplete our bodies of magnesium is by enjoying that fine glass of wine (or any alcohol, for that matter). Although you may not want to read this, it's true: alcohol leeches magnesium from our bodies in a dose-dependent way—meaning that the more alcohol consumed, the more depletion of magnesium.

It's important to choose the right magnesium supplement. Magnesium oxide is not recommended, because it's poorly absorbed. Magnesium carbonate and magnesium citrate can help promote healthy bowels by improving constipation. But when it's used in higher doses, it can cause loose stools. It should not be used if you have a tendency to experience diarrhea. Magnesium glycinate, or magnesium amino acid chelate, is usually more expensive than other forms of magnesium, but it is absorbed more easily and has fewer side effects on your bowels. Magnesium dosing varies greatly depending on the condition being targeted, so you'll want to consult with a health care professional to determine the right dose for you.

### 5. VITAMIN C

This water-soluble vitamin is a potent antioxidant that can protect our bodies from free radical damage due to exposure to toxins and environmental pollutants. What makes this a significant quality is that it can prevent LDL ("bad") cholesterol from becoming stiff. As a result, it reduces your risk of cardiovascular disease. Vitamin C can also stimulate the growth of white blood cells, which boosts your immune system. However, the jury is still out on whether vitamin C has the ability to prevent or treat the common cold.

Vitamin C is an essential vitamin when it comes to collagen and elastin formation. This means it's important for the health of your skin. Collagen is one of the building blocks for your connective tissue, including skin, blood vessels, bones, tendons, and joints. It's also an important vitamin for healing cuts and wounds, and it helps repair tissue and prevents aging.

The therapeutic dosing of vitamin C is unique to each individual. You will know you have reached your maximum dosage when you begin to experience loose stools. This is called dosing until bowel tolerance. It is prudent to note that oxalate is a by-product of vitamin C. When vitamin C is taken in higher doses, it could increase the risk of getting kidney stones (because it can trigger the transformation of calcium into a solid form from a liquid form). For those of you who may have a predisposition to kidney stones, it may be wise to avoid high doses of vitamin C.

### 6. IRON

We need iron to transport and store oxygen. But iron plays a vital role in how our bodies produce energy too. Therefore, low levels of iron can cause fatigue, lethargy, and dizziness. A deficiency in iron can also be a culprit in cold hands and feet, brain fog, and reduced immunity. An athlete with low iron levels will experience poor exercise tolerance and decreased fitness because iron is a critical nutrient contributing to hemoglobin—our red blood cells—and red blood cells are needed to carry oxygen throughout our body. Lower iron levels can also impede our thyroid function, lowering our metabolism. Iron is truly all about energy production.

Iron deficiency is also a risk for women in their childbearing years, for pregnant women, and for vegetarians and vegans. Men and postmenopausal women tend not to have iron deficiencies and should avoid iron supplementation unless their iron values have been tested. Having too much iron in your body can cause iron overload, or, as I like to refer to it, rust the body from the inside out.

Supplementing with iron can be difficult because of its side effects, which include constipation and nausea. Many of us shy away from taking iron supplements for this reason.

The strongest forms of iron are ferrous sulfate, fumarate, and gluconate, and they're available by prescription to correct iron-deficiency anemia. However, these are usually the forms that cause negative side effects. But you can get gentler forms of iron without prescription that are non-constipating, and these include heme and nonheme forms.

Heme iron is absorbed more easily than nonheme iron but may not be an appropriate choice if you're a vegetarian because it comes from an animal source. Be smart when supplementing with iron: too much of a good thing may not, in fact, be good at all. Check with your health care professional first.

### 7. ALPHA LIPOIC ACID

We do everything we can to prevent premature aging. This includes taking antioxidants to prevent the growth of oxidants. We eat deep-colored fruits and vegetables and/or supplement with vitamins A, C, and E; selenium; glutathione; and coenzyme Q10. Yes, they're all vital antioxidants.

So why is alpha lipoic acid on this list if it's just another antioxidant?

Well, it's so much more than just that. Alpha lipoic acid is considered the "universal antioxidant." Unlike other antioxidants, which are either water-soluble and work outside of the cell or are fat-soluble and work inside the cell, alpha lipoic acid is both water- and fat-soluble and therefore works both inside and outside of the cell. Alpha lipoic acid is a powerful antioxidant on its own, but it also has the capacity to bolster other antioxidants by enabling them to be more effectively used by the body. All these properties play a role in slowing down the aging process and promoting better overall health.

Alpha lipoic acid is also vital to balancing your sugar levels, and it also helps lower your cholesterol, helps prevent cataracts and glaucoma, and helps improve your heart health.

Alpha lipoic acid is found in some foods, including red meat, green vegetables, and bran. For each 3-ounce (90 g) serving, beef kidney has 32 mcg of alpha lipoic acid, beef heart has 19 mcg, and beef liver has 14 mcg. One cup (250 mL) of raw spinach has 5 mcg, and the same amount of rice bran has 11 mcg. So supplementation with alpha lipoic acid would be a more convenient way to take in the recommended daily dose—unless, of course, you're prepared to eat 3 lb (1.5 kg) of beef kidneys every day. Alpha lipoic acid commonly comes in 25 mg and 50 mg doses.

### 8. B VITAMINS

The B vitamins are essential for metabolism because they transform our proteins, fats, and carbohydrates into energy. They play a role in our nervous system health, supporting our hormone function and blood cell formation. You need them for maintaining energy during periods of chronic stress and to help support the adrenal glands, which secrete cortisol—your "stress response" hormone.

Vitamins $B_6$ and $B_{12}$ and folic acid prevent inflammation in blood vessels by reducing the protein homocysteine. Vitamin $B_6$ is central in enhancing many neurotransmitter pathways as well. These include serotonin (your "happy hormone" pathway), epinephrine, and norepinephrine. You also need folic acid to produce S-adenosyl methionine (SAMe), which is an important nutrient for brain health. SAMes do a number of good things: they help manage your mood, help make myelin (our nerve cell insulator), and help produce melatonin (discussed in Chapter 8).

A deficiency in B vitamins can result in fatigue; pins and needles sensations in the hands and feet; skin issues, including eczema; irritability; and flat affect (an absence or much-reduced emotional response). It is quite easy to become depleted in B vitamins due to lifestyle factors. Alcohol, refined sugars, nicotine, caffeine, and the oral contraceptive pill all deplete our bodies of the B-complex of vitamins.

But you can build up your B vitamin levels through your diet. Food sources of B vitamins include animal protein (red meat, poultry, and eggs), whole grains, dark-green leafy vegetables, nuts, and some fruits (such as bananas, watermelon, and grapefruit). Bananas and watermelon, however, are not recommended because they are high in sugar and can promote glycation. To augment the B vitamins you consume in your diet, you can supplement with a high-quality B-complex vitamin that includes vitamin $B_{12}$ in a methylcobalamin form and folic acid as 5-methyltetrahydrofolate (5-MTFH).

### 9. PROBIOTICS

Our digestive health plays an important role in our overall health. When our digestion isn't working optimally, toxins can be reabsorbed into the body instead of being eliminated. This causes poor intestinal integrity and discomfort, such as gas, bloating, and digestive upset.

Our bodies then mount an inflammatory response to the toxin overload, thereby disrupting our immune system health. This is where probiotics come into play.

Probiotics contain the same beneficial bacteria that are found in the digestive tract—what some call good gut bacteria. They help strengthen our digestive system, but they also play a big role in our overall health, so they should be an integral part of an everyday supplement protocol.

Probiotics consist mainly of *Lactobacillus acidophilus, Lactobacillus bulgaricus,* and *Bifidobacterium bifidum,* but there are many more. They have anti-inflammatory, anti-pathogenic, and antiallergenic properties and can be used both to prevent disease and to reduce and cure it. There have been many studies showing the benefits of probiotics for digestive health.

There is no recommended daily intake for probiotics, but good food sources include buttermilk, kefir, miso, tempeh, yogurt, and other fermented foods. Supplements are generally higher in potency and are used for therapeutic purposes in treating irritable bowel syndrome, Crohn's disease, colitis, and many other inflammatory conditions.

There are many appealing benefits to taking probiotics, such as strengthening your immune system, replacing good bacteria after a course of antibiotics, and contributing to your general health and well-being.

### 10. AMINO ACIDS

There are three amino acids: arginine, glutamine, and glycine.

*Arginine* Arginine is a semi-essential amino acid. Usually, your body can synthesize enough arginine to meet its physiological demands—although during periods of stress, the synthesis slows. This is worrisome because arginine is responsible for facilitating wound healing and promoting the secretion of key hormones, such as insulin and glucagons. Arginine is also the precursor for synthesizing nitric oxide, which is responsible for decreasing inflammation and oxidation and for boosting the capacity of your blood flow. Arginine also has the ability to prevent blood clotting and "stickiness."

We also know higher doses of arginine can boost growth hormone levels, leading to stronger bones, muscle building, and tissue repair. Growth hormone is responsible for building your muscle, bone, and tissue. It's been proven that exercise and good sleep hygiene help increase growth hormone levels.

Some examples of arginine-rich foods are red meat, fish, poultry, grains, nuts, seeds, and dairy products. If you suffer from cold sores, take note: if you consume a diet rich in arginine foods, you increase your risk of cold sores. Herpes simplex virus 1 (HSV-1) uses arginine to replicate itself. It is not recommended to use arginine supplementation if you have an increased risk for cold sores.

*Glutamine* Glutamine is the amino acid responsible for providing the energy to the cells that line your digestive tract. Its main job is to repair the digestive tract, especially after periods of physical stress. It has the ability to aid in muscle recovery along with arginine and can be prescribed to decrease alcohol cravings. Glutamine, along with glycine, is required to build the most powerful antioxidant of all: glutathione.

The glutamine amino acid has the capacity to support the synthesis of proteins, which are the building blocks of health. This, in turn, leads to muscle building and storing glycogen as energy in the muscle. Glutamine also stabilizes our blood sugar levels by moderating insulin levels and can improve our brain function by providing glucose to brain cells. This gives our brain cells the energy they need to promote alertness, focus, and memory.

Glutamine is present in a number of foods, such as beans, red meat, fish, poultry, and dairy products, but it's hard to achieve therapeutic dosing from food alone. The recommended starting dose is 5 g, and to achieve that, you'll need to look at supplementation.

*Glycine* The smallest and "simplest" of all the amino acids, glycine is needed to metabolize glucose, release growth hormone, for normal cell growth, for cognitive support, and for improved mood. Glycine is also useful for detoxifying many toxic influences, such as exposure to jet fuel, carpeting, dry cleaning solvents, and gasoline.

In fact, taking a dose of glycine can help decrease acute anxiety symptoms as well as decrease symptoms associated with hypoglycemia.

As with most amino acids, food sources of glycine are beans, red meat, fish, poultry, and dairy products. Glycine supplementation is a sweet-tasting powder that is most commonly dosed under the tongue for full absorption. A starting dose of glycine would be 4 g.

These are my top ten supplements, but there are many others that you can use to achieve Peak Health. They include botanical medicines for managing estrogen, progesterone, and testosterone; insulin balancers, such as specific vitamins and minerals; and adrenal glandulars that can help you achieve a balanced cortisol level and improve your energy and stamina.

Finally, please understand that the word *natural* does not mean "safe." It is always a good idea to consult with a naturopathic or osteopathic doctor in order to come up with a plan that will be effective with few or no side effects.

### ONE FOR THE TELOMERES

Dr. Burns and I agree that we could highlight an eleventh supplement for our Top Ten list. This would be glutathione, which we mentioned above, under glutamine.

Ever since we began to measure micronutrient levels when we were testing telomere length in white blood cells, we noticed that many patients who have very short telomeres (in the lowest 25th percentile of the general population) showed a dramatic deficiency in this antioxidant.

Glutathione is not as well known as its other antioxidant family members, such as vitamins C and E. It's also not classed as an "essential nutrient" because it doesn't have to be obtained through food sources; the body can synthesize it from amino acids (cysteine, glutamic acid, and glycine). However, glutathione is critical in all cellular function and is found in all cells. Most people who eat a balanced diet should not be deficient in glutathione, but we're noting a deficiency even in patients who do eat well.

Contrary to popular belief, glutathione can't be absorbed in oral form; it must be administered by nebulizer (inhaled) or intravenously. Alternatively, a product called NAC (N-acetylcysteine) can be ingested to help the body produce more glutathione. If you have very short telomeres, you should talk to your health professional about glutathione.

## Replacing Meals with Protein Shakes

I'm not an advocate of drinking your breakfast, lunch, or dinner. But a protein shake can augment your diet on those days you'd otherwise skip a meal. And enhancing a healthy diet with extra protein can facilitate building muscle mass. Soy powder, whey protein, and yellow pea protein are three choices for adding extra protein to your diet. Consult with your health professional to make sure this kind of supplementation is for you.

Soy powder is a good source of folate, and it is very low in saturated fat and cholesterol. In fact, soy provides important isoflavones, which can reduce cholesterol by as much as 35 per cent and also help prevent cancer. Genistein is one of these isoflavones, and it's been shown to be an anti-cancer agent. But also know that there is evidence that taking soy-based products regularly can stun the function of your thyroid gland, making it sluggish. Soy powder's protein concentration is lower than whey protein.

The most popular protein supplement is whey protein because it is highly digestible and because of its high concentration of amino acids. It is also known to help prevent diabetes, heart disease, and age-related bone loss.

If you're allergic to whey, you can choose yellow pea protein, which is a nonanimal source of protein. This protein is also generally hypoallergenic, and most people find it easily digestible. It has an excellent essential amino acid profile.

## Beware the "Magic Bullets"

This book wouldn't be complete if we didn't explore supplements that make claims to promote the production of the enzyme telomerase and the re-lengthening of telomeres. These claims are often couched in language sanctioned by the Food and Drug Administration.

Some companies claim that silymarin from milk thistle has a positive effect on stimulating the telomerase enzyme; others selling telomere-lengthening supplements claim their products include "trade secret" ingredients.

Astragaloside IV, an active ingredient of the Chinese herb astragalus plant, has been shown to have some anti-inflammatory effects and to stimulate telomerase, which is good[10]—except it hasn't been shown to lengthen our telomeres. The same goes for another active ingredient in the plant, cycloastragenol—although some chemists claim that the two compounds are one and the same: astragalocide IV converts into cycloastragenol once digested.

Two "designer products" that keep appearing in Internet searches are TA-65™ and Product B™. The "best documented" of the astragalus formulations is TA-65™—a secret, patented formula. The patented molecule first developed at Geron Corporation in California includes cycloastragenol, astragaloside IV, and several other related molecules. It's also very expensive at $300 per bottle for a one-month supply.

The other designer product is Product B™, based on the antioxidants vitamin C and vitamin E, milk thistle extract (silymarin), and other herbs. One bottle of this product, which is sold by Isagenix, costs $100 for a one-month supply.

So should you take any of these products?

My answer is no. From my vantage point, no research study can be found to demonstrate reasonable evidence that these new "designer" supplements can stimulate telomerase or re-lengthen telomeres in humans.

As well, we should consider the opposite possibility: if these supplements do work, could they also activate cancer cells and cause them to grow aggressively?

Now, if you are thinking that I'm being a typical "medical doctor" who shies away from Eastern medicine, hold on. My colleague Dr. Burns, a naturopathic doctor, agrees with me. At this time, we don't recommend that our patients spend the necessary hundreds of dollars each month taking these new designer supplements. Our Top Ten list is an excellent place to start–for now.

We'll give the final say to Dr. Elizabeth Blackburn, who earned her Nobel Prize researching telomeres and telomerase. When I asked her about these "designer" supplements, she told me she doesn't take them and worries they may activate cancer.

# 7

# The Power of Hormone-Replacement Therapy

THE AIM OF all hormone-replacement therapy (HRT) is to replace only those hormones that have been lost through the aging process. Replacing them to optimal levels increases our stamina and energy, boosts sexual functions, and restores the physical strength and mental acuity we experienced when we were younger. In the quest for our Peak Health, it is crucial to understand the benefits of the therapy. I support the use of bio-identical hormone therapy in menopause and andropause when patients understand there is a net benefit to using them, especially for the many men and women who are experiencing adverse symptoms that affect their quality of life.

How long you undergo HRT depends on why you decided to use hormones in the first place. HRT involves prescription drugs, and it does have an effective therapeutic effect. If you need it, there's no reason to shy away from medication—you just need to be cautious.

Bio-identical hormones have made headlines over the years. Unfortunately, the headlines offer us mixed messages—and in some cases erroneous and dangerous recommendations. Take celebrities like actress Suzanne Somers, who claims that bio-identical hormones are safer than synthetically derived hormones and can be used for all women who have had breast cancer.

I strongly disagree with her assertion.

Hormones are powerful regulators. Some tumors, such as breast or prostate cancer, can be hormonally sensitive, meaning that they feed on certain hormones whether they are synthetically derived or not.

SCIENCE STAT

**Natural Hormones** Decades ago, doctors first prescribed synthetic hormones to women because they were the only ones that could be taken orally. That was then. Today, some physicians still don't know the significant differences between Provera™ and natural progesterone, and between Premarin™ and natural estrogen.

Premarin™ is a blend of estrogens found in the urine of pregnant mares. These estrogens are natural for horses. But we are human, and our estrogen is different. As a result, Premarin™ can cause side effects, such as excessive water retention and headaches.

Provera™ is a synthetic progesterone called progestin and is most often combined with Premarin™ by prescription. These synthetic chemical hormones do not share the same molecular structure as hormones produced naturally. Natural hormones are defined as hormones that are identical in molecular and chemical form to those produced naturally in the body. In other words, they are bio-identical. Natural hormones do not cause side effects in natural doses, and they also do a better job than synthetics of reducing the effects of menopause. It makes sense that your body recognizes and responds better to them. Fortunately, both estrogen and progesterone come in natural forms.

Each of us is unique, and you will hear me say over and over that the use of any health product must be personalized to your individual needs based on genetics and other biochemical tests. When it comes to hormones, we should first identify if we need them, and then monitor their effectiveness if we do. Each of us metabolizes hormones at a different rate. Unlike using synthetic hormones, using bio-identical hormones allows physicians to measure and monitor hormone levels precisely in the bloodstream. As well, physicians can personalize dosage recommendations more easily.

The abuse of hormone-replacement therapy and other anabolic steroids (man-made chemicals that act like hormone-replacement therapies) by professional athletes has resulted in numerous sixty-second sound-bite news stories about the potential dangers of using hormones. But many of these news pieces detract our attention from the many potential benefits. Hormone-replacement therapy remains a form of age-management therapy, and we need to challenge the negative image associated with this treatment.

Although I do not promote the use of hormone-replacement therapy for bulking up or improving professional athletic performance, there is much research suggesting that hormone replacement has benefits for your telomeres and well-being.

These benefits can be enjoyed when doses are natural, non-abusive, and controlled. When monitored, this rational approach to hormone-replacement therapy should cause no significant negative side effects for you. In fact, I can guarantee that it will not create a race of bestially aggressive, over-muscled, and hypersexual humans.

### The Role of Hormones

I like to compare the delicate art of balancing hormones to conducting an orchestra. You have your own orchestra playing within you, and you're producing your own unique music—depending on how you conduct yourself. In the same way, there are many hormones performing different functions in your body, and they all have one common purpose: to keep you alive. Hormone levels vary among individuals, and those variations, in turn, contribute to your individual health and personality.

The word *hormone* comes from the Greek and means to "set in motion" or "to spur on," and this is exactly what our hormones do. They set in motion the many biological functions that keep us healthy. And because our biological system is designed to keep the body and mind in balance, we must keep a careful balance of the eighty different hormones that work together continuously to maintain optimal health.

Your endocrine system is responsible for balancing hormones. It's made up of glands that produce and send hormones, acting like messengers to all parts of your body. Your endocrine system regulates processes such as metabolism, cell growth, cell aging, and cell death. The target cells receive the hormone, which triggers the cell to act in a specific way—to synthesize proteins, replicate, or repair itself. All these commands are part of your body's overall response to its environment. As well as controlling the actions of individual cells, hormones also send feedback messages to your different glands, indicating if more or less of any hormone is needed in your body.

### Hormones and Anti-Aging

In the world of anti-aging, we talk about hormones all the time. Why? Because they change our cell metabolism to support growth and rejuvenation, which produces the effects of youthfulness. Scientists are now also learning we can use our hormones to activate telomerase—our biological regulator of telomere length.

Before we get into the details about specific hormones, I want to explore what happens to our hormones as we age. Most hormone secretion is plentiful when we're children and growing, but those levels decrease as we get older. That secretion affects your telomeres as well.

When your hormones are plentiful, telomerase replenishes the telomere portion of your DNA. But when the hormone levels drop off, the telomerase protein also slows its repair action on your telomeres, causing your cells to age.

The 2009 Nobel Prize in Medicine for discovering telomeres and telomerase created a new field of treatment called telomerase-activating therapy. Hormone replacement is one strategy in this therapy.

Hormone production depends on two mechanisms: genetics and lifestyle. When we begin to experience a lower hormone level, regardless of cause, we are essentially unable to send a strong message to another part of the body, a message that would instruct it to perform an important bodily function. When a hormone reaches a critically low value, we can sense it and we describe it as symptoms. For example, when your estrogen levels are low, most women begin to experience hot flashes and night sweats. Men often complain of lower libido when their testosterone levels drop. When you are stressed, your cortisol hormone spikes and you can feel your heart racing.

You can think about how hormone production works in the same way as good personal relationships work. When we stop talking to each other, or when we communicate badly, the quality of the relationship usually suffers and discord results.

Genetically, hormone levels change over time; this function is programmed in our DNA. But when we add environmental stimuli and cause changes to this blueprint program, our ability to precisely balance our hormone levels can be easily lost, and the result is premature aging of our tissues. Indeed, achieving hormonal balance is critical to maximizing our telomere function.

For example, some research suggests that enhancing our well-being through anabolic hormones (testosterone and dehydroepiandrosterone, or DHEA)[1] and growth hormone[2] can be linked to longer telomere length. I discuss these hormones later in this chapter.

Although scientists focus on more than eighty hormones in their research, I want to focus on just a few hormones—in particular, the ones that we can easily measure and manipulate medically. Some of these hormone levels drop as we grow older because of genetics, and others are sensitive to lifestyle factors, such as diet, sleep, and exercise. Together, these hormones affect the rate of glycation, oxidation, and inflammation—which you'll recall are the three major processes of aging and telomere shortening. These hormones are:

- Estrogen
- Progesterone
- Testosterone
- DHEA
- Growth hormone
- Insulin (discussed in Chapter 4)
- Melatonin (discussed in Chapter 8)
- Cortisol (discussed in Chapter 9)
- Serotonin (discussed in Chapter 9)

### HORMONE-REPLACEMENT THERAPY PROS AND CONS

There are a number of valid benefits from hormone-replacement therapy. These include:

- A reduction in the incidence of bone fractures due to osteoporosis
- A reduction in the risk of colon cancer
- A boost to your libido and sexual function

- An improvement in the texture of your skin
- An improvement in your sense of emotional well-being—relieving anxiety and depression-like symptoms[3]

However, there are also a number of adverse side effects, and these include:

- Blood clots—especially for those with genetically based clotting defects
- High blood pressure
- Headache
- Gall bladder and liver dysfunction
- Excessive water retention

## Replacing the Sex Hormones: Estrogen, Progesterone, Testosterone, and DHEA

Estrogen, progesterone, testosterone, and DHEA are our sex steroids, or anabolic hormones, and they deserve the first look. After all, their critical job is to make sure we can reproduce. Once we are past our prime to bear children, their levels all decline. This period of time is known as menopause for women. In men, it is called andropause. Women often feel more dramatic symptoms than men. Nevertheless, both sexes dislike the symptoms associated with these hormone levels changing. These include:

- Sexual dysfunction
- Mood changes
- Altered body shape

Because of these symptoms, many people decide to undergo hormone-replacement therapy. Let's dive into these hormones and understand what can be done to manage menopause and andropause. Note that I use the word *manage* and not *treat*. These are natural phases in our lives, not diseases. Let's go into further detail about replacing each sex hormone.

## ESTROGEN

Estrogen is mainly a female hormone, but it is also found in small amounts in men. In women, estrogen is produced mainly in their ovaries. Each month, estrogen stimulates ovulation. It also helps prepare their uterine lining to receive a fertilized egg. Estrogen is also responsible for physical maturation, including the filling out of the breasts and hips, and the development of the female reproductive organs during puberty. In men, this hormone is responsible for the maturation of sperm.

In women, there are three types of estrogens. Estrone (E1) is the predominant estrogen during menopause; estradiol (E2) is most potent and dominant during a woman's reproductive years; and estriol (E3) is most abundant during pregnancy. All three types of estrogens are synthesized from male sex hormones (testosterone and androstenedione).

Metabolically, estrogen accelerates the burning of fat, increases HDL ("good") cholesterol while decreasing LDL ("bad") cholesterol, and reduces bone resorption, which boosts bone formation. It's no wonder menopausal women notice a 10 to 15 per cent weight increase, a sudden increase in their cholesterol levels, and the development of osteoporosis (bone loss).

Cortisol, our stress-response hormone, is also regulated to some degree by estrogen—which partly explains hot flashes, sweats, and palpitations during menopause. A rush of these adrenaline-like symptoms can be very debilitating during the workday and at night, when you're trying to get a deep, long sleep. Chronically poor sleep can provoke additional hormonal imbalances (reducing your cortisol, growth hormone, and serotonin)—a vicious cycle.

And do you ever wonder why skin looks so much better in non-menopausal women? Estrogen regulates healing through collagen, which is a fibrous connective tissue protein in our skin cell walls. The more collagen produced, the better

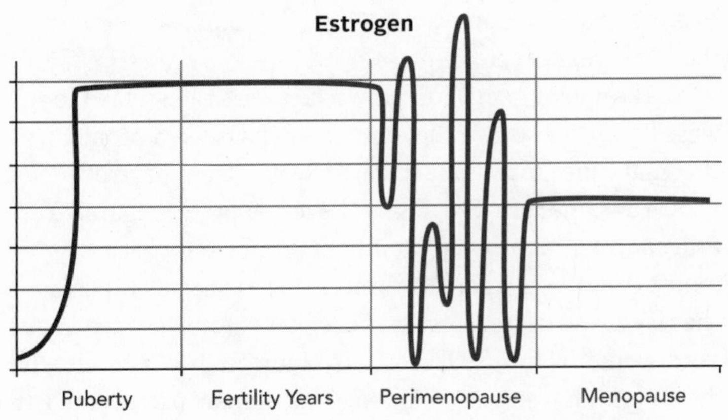


This is a graphic representation of estrogen levels in women from puberty to menopause. In women, estrogen hormone production begins at puberty and is marked with the onset of menstrual cycles, which continue during the reproductive years. Around the age of 45 to 50, estrogen levels decline, marking the onset of menopause.

our skin looks. Don't we often say that pregnant women have such a beautiful glow to their faces? Now you know why.

Equally important, many scientists believe estrogen has an anti-inflammatory effect, which helps to slow down the aging process.

*Estrogen-Replacement Therapy (ERT)* The benefits of estrogen-replacement therapy have been hotly debated for over twenty-five years. In the 1980s, ERT was all the rage until the Women's Health Initiative (WHI) study showed there were risks connected to hormone therapy that could outweigh their benefits.[4]

This study included clinical trials that assessed the effects of postmenopausal hormone therapy, and diet modification, along with calcium and vitamin D supplements, against the risks of heart disease, fractures, and breast and colorectal cancers.

At one time, it was believed that taking hormones decreased cardiovascular disease. In fact, the WHI study showed in 2002 that the opposite was true in some women during their first year of estrogen use.

So the question millions of women were then forced to ask was: "If I use ERT to relieve my menopause symptoms, will that make me more likely to have a heart attack or stroke?"

At the time, ERT was also believed to protect against certain cancers. But in later studies, it turned out that ERT promoted the growth of breast tumors. To be exact, ERT increased the risk by eight new breast cancer cases for every ten thousand women who used ERT daily compared to women who didn't take hormones. The longer a woman participated in ERT, the more it seemed her cancer risk increased. The good news is that, within three years of stopping ERT, the risk decreased to that of the general non-ERT population.

What about ovarian cancer risk? Although deadly, this type of cancer is rare, and the WHI found that there may be a slightly higher risk of developing ovarian cancer over time due to ERT.

In 2010, the WHI confirmed that a woman's risk of heart disease more than doubles within the first two years of taking combination-hormone replacement (estrogen with progesterone) within ten years from the onset of menopause.

Some of the WHI study results and findings were controversial because these trials involved a large number of women over a long period of time. Critics felt that the age of the women in the studies skewed older (inferring the presence of more chronic disease conditions). Others argued that the hormones used were synthetic—Premarin™ and Provera™—rather than bio-identical, or natural, hormones.

Doctors now have a better understanding of the importance of balancing and replacing all the body's hormones, not just one. Using progesterone together with estrogen, for example, has been found to be critically important in preventing estrogen replacement–related uterine cancer.

There is also some good news. Scientists now believe that estrogen can slow the decline of cells that age quickly in post-menopausal women who carry the ApoE4 gene (a genetic risk for Alzheimer's disease).[5] Once halting a women's hormone treatment regimen, these genetic carriers exhibited much faster telomere shortening. This study supports the use of estrogen replacement for women at a high risk for developing age-onset dementia.[6]

The WHI Study also showed that ERT reduces the risk of colon cancer by about 40 per cent. The risk reduction decreases significantly once ERT is withdrawn for approximately two years.

A study on estrogen-sensitive tissues shows that telomerase is activated by the treatment of estrogen.[7] This may be one factor that contributes to women living longer than men. These findings may help us discover the mechanisms of estrogen's effect on telomerase activity and the role of this hormone on cellular aging, which occurs after menopause, as well as its effect on inducing cancer. This is a very important connection because the mechanism of estrogen may prevent some of the chronic diseases, such as osteoporosis and colon cancer, because of its ability to "turn on" telomerase, keeping telomeres long and healthy.

Women have been prescribed estrogen to relieve their menopausal symptoms for decades. This remains true today. It does work, and from my perspective, it remains safe in most women so long as there is appropriate prescreening as well as ongoing screening during and after using ERT.

All this to say, estrogen replacement must be used with caution. I do not shy away from prescribing estrogen. However, each woman has her unique set of risks and benefits and these need to be assessed on an individual basis to determine if estrogen-replacement therapy is right for her.

### A STORY ABOUT A DISTRESSED WOMAN

When you think of menopause, you might envision a woman with sweat pouring down her face or flipping the bedcovers on and off all night, irritating her sleeping partner to no end. These behaviors are caused by hot flashes. Although these scenarios appear to be generally benign, I'd like to share a story of a client whose menopausal symptoms dramatically altered her life. She first showed up at my office in tears, telling me she had been unable to sleep for three months. She could not concentrate at work, she was uncharacteristically panicky, and she was yelling at her staff and family. She felt her job as a senior vice president at a Fortune 500 company was too stressful and wanted to submit her resignation letter the next day. The only problem was that she was feeling too exhausted to draft this letter. On further discussion, it became apparent to me that she had been experiencing significant side effects of menopause—hot flashes, night sweats, mood swings, and insomnia. I urged her to take a few days off and get some sleep with the help of a medication, and she started estrogen-replacement therapy. In a few days, she began to feel better and decided to hold off on resigning.

Thank goodness. Hormone-replacement therapy prevented her from making an irrational decision and saved her job.

---

I also recommend herbal, or "natural," products in my practice to help my patients relieve their symptoms of menopause—especially to reduce hot flashes and sweats. In particular, soy-based phytohormones (plant estrogens) and herbal products (black cohosh and red clover) have been shown to have a positive effect. However, the dose must be high enough to be therapeutic and it must be used regularly. These natural chemicals have similar structures to estrogen molecules. Although the analogy is not technically correct, I'd like to say that they behave like "fake" estrogen to the body's estrogen receptors. You can fit the key into the lock but it doesn't unlock the door to unleash some of the serious side effects I noted with ERT.

I do agree with my naturopathic colleagues: if there is a risk of serious side effects in using HRT, it's best to avoid all hormone replacements, even natural products. This is especially true with women who have had breast cancer or have a risk of blood clotting.

### PROGESTERONE

Progesterone is produced mainly in a woman's corpus luteum (the follicle that ruptures during ovulation) and in the placenta during pregnancy. It also plays a role in determining whether a menstruation happens or not. If there is no fertilized egg, progesterone is not needed to help prepare the lining of the uterus to receive the egg. As a result, progesterone levels drop, triggering the sloughing off of the lining of the uterus, which can also cause some women to experience

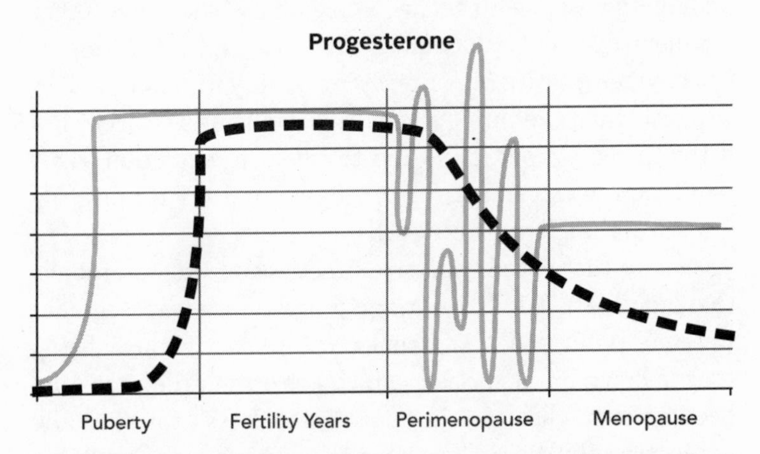


This is a graphic representation of relative progesterone levels in women from puberty to menopause. The dotted line represents progesterone secretion throughout a woman's reproductive life cycle. The background lines represent estrogen levels over the same time period. Progesterone declines before estrogen.

premenstrual syndrome (PMS) symptoms. PMS symptoms can include bloating, water retention, moodiness, and irritability.

*Progesterone-Replacement Therapy* In a woman's body, estrogen works very closely with progesterone to regulate the menstrual cycle. Progesterone is produced every month in a cyclical way. The overall level of progesterone begins to drop when a woman is in her early forties. By the time menopause occurs, the amount of progesterone in her body fluctuates greatly, creating PMS symptoms. In fact, most of the menopausal discomfort women feel is due to the lack of balance between estrogen and progesterone levels.

Although estrogen alleviates the symptoms of menopause, progesterone also helps eliminate hot flashes and mood swings.

Earlier, I discussed how important progesterone replacement therapy is in reducing the risk of uterine cancer when undergoing estrogen therapy. Researchers have shown that telomerase activity fluctuates with progesterone levels. It peaks when the uterus is ready for embryo implant and is weakest when the lining sheds. This confirms a relationship between sex steroid levels and telomerase regulation.[8] This regulation is important in turning on and off the healthy growth of the uterine lining cells.

One of the more impressive functions of progesterone is that it builds bones. Estrogen and progesterone work together to remodel our bones. Old bone cells are continuously being broken down by osteoclasts, which I've often called Pac-Men because they chew at old bone. Osteoblasts stimulate the production of new, healthy bone cells. This bone remodeling occurs throughout our lives.

Estrogen plays a more dominant role in resorption (osteoclastic function), and progesterone plays a larger role in bone formation (osteoblastic function). The fact is, if you're a postmenopausal woman, you will lose between 2 and 4 per cent of your bone mass each year. This happens via the imbalance between the resorption and formation of bone and is caused by changes in hormone levels with age. During early menopause, your progesterone levels drop much faster than estrogen levels, which causes a higher resorption rate than bone-formation rate.

### TESTOSTERONE

Testosterone is mostly responsible for men's sexual development during puberty, and it is critical in maintaining erectile function and libido. This hormone also declines with age. By the time a man reaches 80, his testosterone levels have dropped to only 20 per cent of what they were in his twenties.

Testosterone plays an important role in the formation of male organs and in secondary sex characteristics, such as facial and pubic hair growth, increased muscle mass, and the development of a deeper voice. In both sexes, this hormone is considered an anabolic, or bodybuilding, hormone that promotes increased muscle mass. Testosterone binds to receptors in your muscle cells. This increases the rate of nitrogen retention, which stimulates muscle growth while preventing muscle breakdown. Some experts believe that testosterone aids in weight loss because this hormone "burns" fat. As well, testosterone increases your bone density, body hair growth, and oiliness in the skin, which can sometimes lead to acne. We often see this among teenagers of both sexes.

Some studies suggest that higher levels of testosterone in men can reduce heart disease because more blood flows to the heart muscles.[9] Metabolically, testosterone cuts the rate of glycation by helping your body maintain lean body mass, reduce visceral fat and total cholesterol, and control your blood sugar.

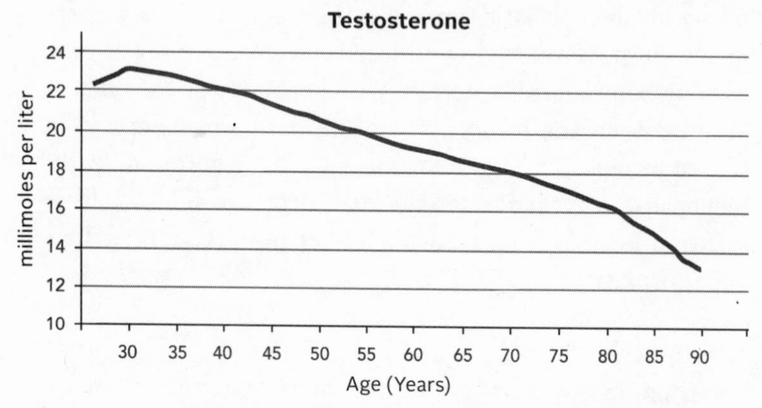


In men, testosterone levels initially increase during puberty, peaking around age 30, and then decline over time. When testosterone levels become low enough, men enter andropause.

There is growing scientific evidence to suggest that healthy levels of testosterone maintain our brain functions, such as memory and attention span, and that lower levels may put us at risk of mental decline.[10]

*Testosterone-Replacement Therapy (TRT)* Research with testosterone dates back to the mid-1800s, when German physiologist Arnold Berthold removed the testes from two young roosters and transplanted them into two older roosters. The result? The castrated roosters became lazy and fat. The two older ones became more energetic, and their sex organs (their combs) kept growing.

In 1939, German biochemist Adolf Butenandt won a Nobel Prize for isolating the testosterone molecule. Some reports claim that the Nazis used testosterone during World War II to make their soldiers more aggressive. In 1945, journalist Paul de Kruif wrote a book called *The Male Hormone*. He referred to testosterone as "medical dynamite" and "sexual TNT." Its benefits have been known for such a long time. The question is, why do men with hormone deficiencies shun it?

The most common reason is the connection between testosterone-replacement therapy (TRT) and prostate cancer. Some studies showed that testosterone produces dihydrotestosterone, which speeds the growth of prostate tumors.[11] In contrast, Michael Eisenberg of Stanford University presented a study at a 2013 meeting of the American Urological Association that found that men undergoing TRT had a lower likelihood of developing prostate cancer than those not undergoing TRT.[12]

DEFINITION

**Prostate-Specific Antigen** PSA is a blood measurement used to assess the presence of prostate disease. Both benign and malignant prostate issues, as well as some drugs, can increase PSA levels. Nevertheless, it is a useful tool. A PSA level can be inputted into a prostate risk calculator to improve the predictive ability of a PSA value in determining your risk of prostate cancer. Dr. Robert Nam at Sunnybrook Health Sciences Centre in Toronto, Canada, created one such calculator. You can access it at sunnybrook.ca/content/?page=occ-prostatecalc.

Because young men have higher testosterone levels than older men and because prostate cancer is generally not found in young men, it seems reasonable to believe that testosterone does not cause prostate cancer.

Prostate enlargement is common as men age. And testosterone therapy can also cause the prostate to grow larger, which increases the risk of prostatism. This is a diagnosis of symptoms caused by an enlarged prostate, such as slow urine stream and increased urinary frequency, especially when you're asleep.

The good news is that no study to date has shown that using TRT increases the frequency of prostate cancer, the most common cancer in men. However, it does accelerate prostate cancer growth if cancer is already present. Therefore, routine prostate cancer screening should be mandatory before you decide to undergo TRT and during treatment.

Some men have been reluctant to consider TRT because of the nature of andropause, or male menopause. Many people compare andropause to menopause, but there are some big differences. When menopause occurs, women experience a distinct physiological change: their menstrual periods cease. There is no such obvious event in men when andropause has arrived. Another difference is that andropause can start anywhere between the ages of 30 and 50, and women go through menopause almost always in their late forties or early fifties.

Andropause symptoms also appear more gradually, as a progressive decrease in energy, depressed mood, and impaired sexual function. Men will also experience a thinning of their bones, a weakening of their muscle strength, and increased body fat. Research also links testosterone deficiency with hypertension and an increased risk of heart disease.[13]

Because of the link between testosterone, sexuality, and the world's perception of manhood, many men are reluctant to talk about andropause or a possible testosterone deficiency.

In recent years, however, there has been a real surge of testosterone use among men. It seems their concerns about developing prostate cancer have been pushed aside in favor of the positive effects of testosterone replacement. Often, testosterone is combined with growth hormone. This combination boosts the effects of improved muscle and bone mass as well as improved energy.

Although osteoporosis is often considered a "woman's problem," it's a serious concern for men as well. In fact, one in eight men over age 50 will have an osteoporosis-related fracture. One study revealed that testosterone-deficient older men were six times more likely to break a hip during a fall than younger men. When the hormone is replaced to levels you would expect to see in younger men, older men display an increase in bone density, bone formation, and bone minerals—reducing their risk of bone fracture should they fall.[14]

Another study found that men who supplemented with testosterone experienced a significant decrease in their body fat and a change in the overall physical composition of more muscle. What was the change? Mainly, their pot-bellies disappeared.[15] Supplementing with the hormone also improved their blood glucose levels, decreased their blood pressure, and increased their insulin sensitivity.[16]

Scientists once thought that too much testosterone contributed to heart disease in men. The thinking was that, since estrogen helped prevent heart disease in women, testosterone must be the culprit in male heart disease. But sex hormones work differently in men and women, and research links low testosterone levels and heart attacks in men. Dr. Gerald Phillips of Columbia University evaluated blood vessel imaging of fifty-five men. He found lower testosterone levels were present among men who had a higher incidence and severity of coronary artery narrowing. Dr. Phillips also discovered that testosterone appeared to raise the HDL ("good") cholesterol levels and to drop LDL ("bad") and total cholesterol levels.[17] This may be one main reason why testosterone seems to prevent heart disease.

There's another critical benefit of testosterone. Scientists at the Johns Hopkins University first showed that testosterone activates telomerase and lengthens telomeres.[18] They used testosterone hormones to regulate telomerase activity in the stem cells of rat prostates, proving the importance of hormonal regulation of telomerase activity.

Just like estrogen and progesterone, a woman's testosterone levels drop as she ages. If testosterone is not at its optimal level, the female body becomes a bit out of balance. So we shouldn't rule out the importance of adjusting testosterone levels with testosterone-replacement therapy in women. Studies have shown that women who undergo TRT have improved sexual desire.[19] However, clinical trials have not shown strong evidence to suggest that TRT alone is enough to stimulate a woman's sex drive. In my clinical experience, adding an estrogen replacement is critical to improving a woman's sex drive.

Although testosterone can't help rebuild our bones completely, estrogen does have the ability to decrease the destruction of bone in postmenopausal women. Like progesterone, when testosterone is given with estrogen, damaged bones begin to rebuild and strengthen.

But beware of overdosing with testosterone. Doing so would mimic the most common natural cause of higher testosterone levels in women—polycystic ovarian syndrome. These women experience oily skin, male-pattern hair loss, increased body and facial hair growth, a deeper voice, and a change to their body shape.

In conclusion, as in estrogen-replacement therapy, each person has his or her unique set of risks and benefits, and these need to be assessed on an individual basis to determine if testosterone-replacement therapy is right for him or her.

### DHEA

Dehydroepiandrosterone (DHEA), known as the other anabolic hormone, is the most abundant steroid in our bodies. It's mainly a metabolic intermediate. This means that DHEA is involved in the production of testosterone and estrogen, so it can be used as a hormone-replacement solution because it can augment testosterone and estrogen levels.

*DHEA-Replacement Therapy* DHEA is produced naturally in your brain and your adrenal cortex. This abundant hormone is a precursor to the sex hormones I've just discussed. DHEA levels peak at about age 30 and then drop steadily afterward.

CASE STUDY

**A Story about a Hairy Supplement Gone Wrong**

One of my most-often-told patient stories involves the unmonitored use of hormones. One woman, a national television personality, had been a client of mine for many years. Her main desire was to stay healthy and young, a wish of so many women in their mid-forties. For her, being on TV meant that her looks mattered, and so she was always looking for the next potion. Although she was perimenopausal (still having her periods), she was not yet a candidate for estrogen and progesterone therapy. So we balanced her hormones with some effective phytohormones (herbal supplements) and checked her biometrics thoroughly every year. Out of the blue, she called one day and told me that she was "freaking out" because "handfuls of hair" were falling out from her head. She was breaking out with acne all over her face and body. Although she usually took months to finally make her annual appointment, she showed up in twenty-four hours and I ordered a full set of lab tests.

To my horror, her DHEA level was at 33. Normal female levels should be no higher than 10. As well, her testosterone was twice the normal level. I called her and pointed out that she was either taking hormones or had an adrenal tumor until proven otherwise. She denied using anything other than what was prescribed by me, but on further prompting, she told me about a doctor she'd done a story on who gave her some "supplements" to help with her menopausal symptoms. The "supplement" turned out to be DHEA. Unbeknownst to her, she had been consuming a man's dose of DHEA and had been doing so for six months—unsupervised!

Thankfully, with the withdrawal of DHEA, the symptoms subsided over three to six months. Today, television viewers still enjoy seeing her at work—with a full head of hair and free of acne.

DHEA products claim to energize your body; restore sex drive; improve memory; lower cholesterol; fight obesity, heart disease, and stress; and strengthen your immune system. People who take DHEA say they look and feel years younger.[20]

This hormone is popular with many who want to experience the effects of HRT without taking estrogen or testosterone. Because DHEA is a building block of these hormones, the thinking is that the body won't become overloaded with estrogen or testosterone and that it will only make what it needs. However, this theory is not true.

Using inappropriate doses of DHEA without medical supervision is as dangerous as taking estrogen and testosterone. Indeed, with the use of DHEA, testosterone and estrogen levels rise, and so do the risks and benefits. So DHEA levels must be monitored. Again, speak with a physician who specializes in hormone replacement before supplementing with DHEA.

This hormone is easily available as a natural supplement in the United States. At one point, Health Canada banned the sale of DHEA altogether. Today, through physicians, DHEA can be prescribed. You can find hundreds of ads for DHEA on the Internet, but it's better to get it from a reputable health store in the United States or a physician in Canada so you're sure of getting the real thing.

*Growth-Hormone Therapy (hGH)* There has been a lot of media coverage of the use of anabolic steroids and growth hormone by professional athletes. I know some sports medicine physicians who inject growth hormone right into injury sites to promote quicker healing of injured ligaments, tendons, and muscles.

In the anti-aging world, hGH is popular because it improves your vitality. This hormone has powerful effects on

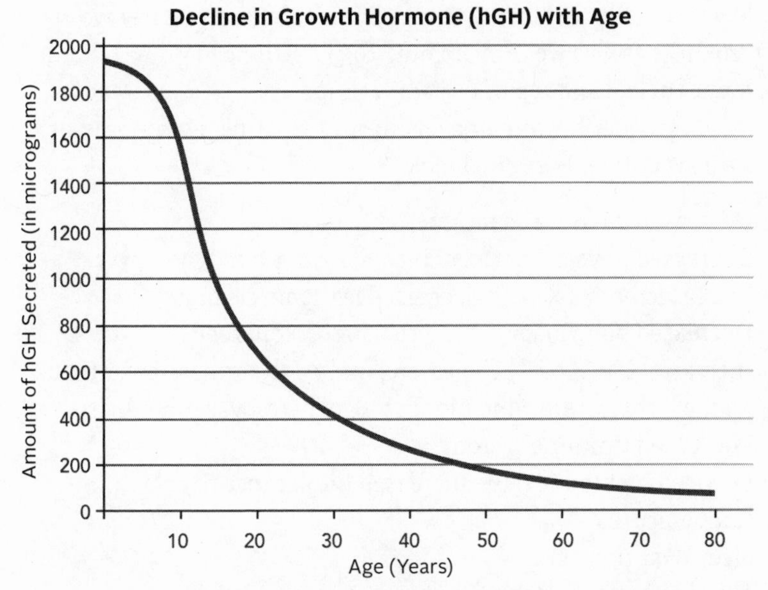


Growth hormone secretion declines continuously with age. By age 80, our hGH levels drop to only 5% of what they were at age 20.

healing and regeneration, and it is responsible for growth and repair of all tissues in your body. It plays a significant role in repairing muscles and remodeling bones as well as in managing our visceral (internal organ) fatty deposits and in cholesterol profiles.

Growth hormone is produced by the pituitary gland in our brain. Most of us are born with high levels of growth hormone. Children with genetically low growth hormone experience stunted growth, and hormone replacement is used to ensure that they can grow. In contrast, children and young adults who have too much growth hormone develop a condition called acromegaly, which many believe President Abraham Lincoln had, and he grew to be six-foot-four.

As we age, our growth hormone level drops. By the time we are in our sixties, our level of growth hormone is much lower than during our twenties. But after puberty, after we've grown as tall as we're ever going to, we still need growth hormone to heal and repair our adult tissues.

As growth hormone levels drop over time, symptoms of aging increase. These include:

- Decreased energy and fatigue
- Decreased physical performance, stamina, and recovery
- Increased body fat with decreased lean muscle mass
- Decreased bone mass–rising risk for osteoporosis
- Increased LDL ("bad") cholesterol and triglycerides, lower HDL ("good") cholesterol, and higher risk of cardiovascular disease
- Weakened immune system
- Reduced metabolic rate–increased intolerance to cold temperatures
- Sleep dysfunction
- Mood alteration–decreased libido, anxiety, and depression
- Impaired memory and ability to concentrate

Although our pituitary gland releases small amounts of hGH throughout the day, most of our hGH is secreted at night, peaking one to two hours after the onset of deep, rapid eye movement (REM) sleep. Therefore, it's essential to get enough good sleep to ensure the optimal secretion of growth hormone every day. Strenuous exercise every day and sufficient intake of high-quality protein are also critical for optimizing your hGH secretion.

The ability of hGH to improve and reverse the age-related side effects of physical stamina and mental acuity means that hGH replacement is popular not only among athletes but also among baby boomers. The problem is, it's not easy to access growth hormone without a prescription. And in reality, it's a gray zone for many doctors. Presently, the conventional wisdom is to prescribe growth hormone replacement only to people who have adult onset growth hormone deficiency due to illness, such as a brain tumor, for example—and not for reversing the natural effects of aging.

Growth hormone is given by injection on a daily basis. It is critically important to monitor growth hormone levels to reduce the side-effect risks. To begin with, growth hormone affects how our body uses insulin, so monitoring our blood sugar levels is important. Before undergoing treatment, a patient should be carefully screened for cancer because this hormone promotes the growth of all tissues—including tumors. If a patient uses too much hormone, they will notice muscle and joint pain, fluid retention, and the effects of carpal tunnel syndrome (pain and numbness in the hands due to a swelling of the wrist ligaments).

That said, growth-hormone-replacement therapy is picking up steam all across the developed world. If you decide to try it, do so with a trained physician who understands how to carefully measure the correct dose for you and will monitor your side effects on a regular basis—not just once a year! You can also improve your growth hormone levels by staying at a healthy weight, exercising regularly, sleeping well, and eating enough protein. This sounds like a repeating theme, doesn't it?

Tissue repair ideally happens when we're sleeping, and my patients who show chronic sleep deprivation (for example, sleep apnea) have relatively lower growth hormone levels. Without adequate protein, no hormone can be produced. Arginine, lysine, and ornithine amino acids, derived from protein sources, are critical to growth hormone production. If you don't consume sufficient protein in your diet, it's possible to boost your growth hormone levels by supplementing with these three amino acids.

Before we had a better understanding of how telomerase works, I was curious to know more about the mechanism of growth hormone therapy in healing and repairing injuries, and why it has been so popular among athletes—amateur, professional, and the "young" 60+ athletes.

A team of researchers at La Paz University showed that growth hormone directly activates telomerase.[21] This is why hGH has become popular among medical sports specialists. Perhaps by lengthening the telomeres of tissues, they are able to keep them robust and strong, making them more resistant to injury.

SCIENCE STAT

**Doping** We know that many professional athletes use hormones to boost their performance. The most notorious and successful at this is cyclist Lance Armstrong. His use of testosterone and growth hormone therapy with blood doping is banned in competitive sports. However, we can't dispute the proven effects with his significant wins on the cycling circuit. Off the circuit, I wonder out loud if his doping flared the aggressiveness of his testicular cancer.

### Hormones and Telomeres: On the Way to Peak Health

There is no doubt in my mind that supplementing with hormones will have an overall beneficial effect on well-being. There will be continuing debate as to the role of hormone replacement therapy in maintaining telomere health. I'll end this chapter with a conversation that I had with a telomere scientist who asked me a question as I was driving to our speaking event.

"Are you in menopause yet?" she asked.

"I'm not sure, as I'm on hormone replacement to prevent the onset of menopausal symptoms," I replied.

"That's good—it's not published yet, but estrogen does help maintain telomere length, but only if you start it very shortly after menopause begins—otherwise it's not so useful."

SCREEN YOUR RISKS

**Before Trying Hormone-Replacement Therapy**

Despite the benefits of hormone supplementation of hormones that drop with age, there are real health risks too. You should speak with an experienced medical professional to help you decide if hormone replacement is right for you. Before you begin, you'll want to undergo an extensive set of screening diagnostics to ensure that the hormones you're considering will not aggravate any preexisting condition. This should include performing a set of screening blood tests to ensure that your liver and kidneys are able to process the hormones as they are metabolized in your body. Once you begin, hormone levels must be monitored regularly—just one imbalanced hormone can affect the others. Always know your baseline hormone levels and determine what your goals are. Recheck hormone levels at least twice per year. Be vigilant at all times, looking for adverse side effects. Exclude other causes for troubling menopausal or andropausal symptoms. Imaging may be warranted. This could include a breast or prostate ultrasound, a mammogram and breast or prostate MRI, an abdominal and pelvic ultrasound, and cardiac scans.

# 8

# Age-Proof Your Lifestyle

## Small Choices Make a Big Difference

Your telomeres are telling on you; it's time to age-proof your lifestyle. We've known for years that not getting enough sleep, not exercising enough, and using recreational drugs can impact your health in a bad way.

But now we can see the effects of these factors in the length of our telomeres.

## Exercise

It's well documented in scientific literature that regular exercise can reduce the onset of age-related cardiovascular disease, hypertension, and diabetes. But until recently, there was very little molecular evidence to link exercise with a reduction in your overall risk for disease and your chances of living a longer life. Telomere length may provide that missing thread of molecular evidence.

Longer telomeres have been found in people who are more physically active,[1] and this connection has been the focus of

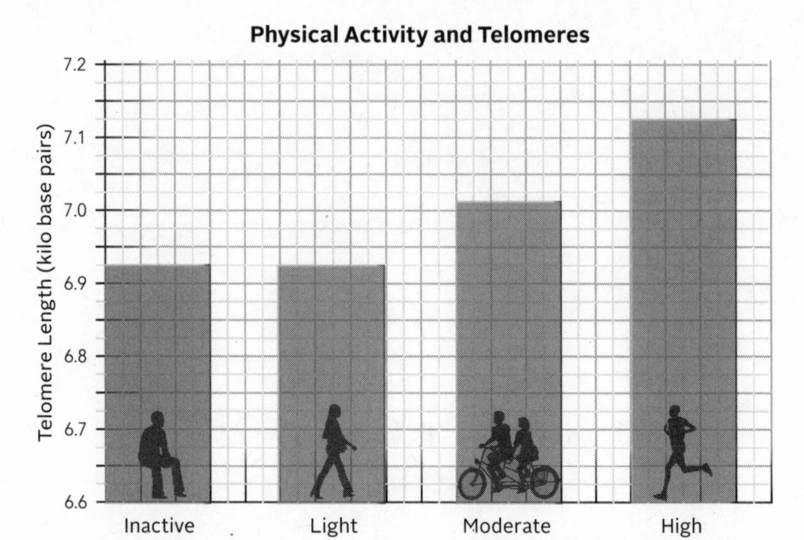

The more active a person is, the longer their telomeres are.

a number of recent studies. "Physical Activity and Telomere Biology: Exploring the Link with Aging-Related Disease Prevention," published in the *Journal of Aging Research* in 2011, is just one example. Because our telomere length is also connected to a reduced risk of disease and improved longevity, it appears that levels of physical activity do have an impact.

Another study looked at the different activity levels of more than one thousand sets of identical twins. It was found that the twin who was more active had longer telomeres, even after adjusting for such factors as age, smoking, and body mass index (BMI). The study went on to suggest that inactive subjects may be biologically older by more than a decade compared to the more active subjects.[2]

It's important to understand that fitness and exercise are not necessarily the same thing. Many people exercise but they're not fit and healthy. Their exercise is inadequate and doesn't stimulate the metabolic changes they need to maintain or improve the length of their telomeres.

**Twins, Activity, and Telomeres**

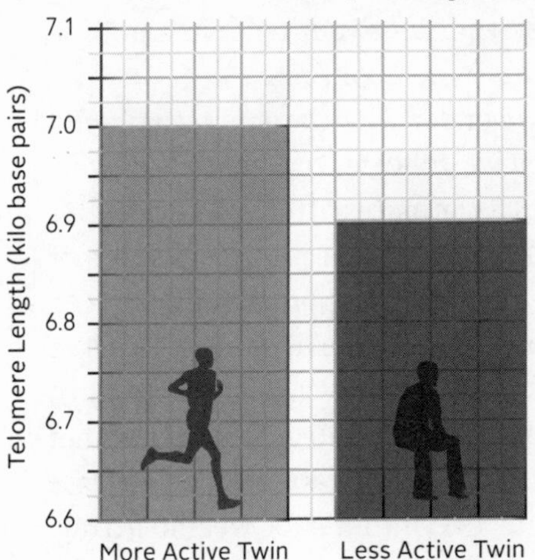


In identical twins, both individuals begin life with the same number of telomeres. Environmental factors, such as activity levels, affect each twin independently of their genetics—proving lifestyle can modify one's genetic predisposition.

Exercise reduces the effects of aging by:

- Lowering excessive blood glucose and insulin, which prevents glycation
- Elevating growth hormone, testosterone, and serotonin, which counteracts aging hormonal imbalance
- Lowering blood pressure and cholesterol levels, which decreases cardiovascular inflammation
- Inducing natural body antioxidants, which reduces oxidative stress

### AEROBIC AND ANAEROBIC EXERCISE

The main types of exercise—aerobic and anaerobic—both affect key aging-related hormones in all of us. One doesn't replace the other.

Aerobic exercise includes activities such as jogging, walking, and going up and down stairs. These types of activities improve your cardiac fitness and pulmonary function—they're very important for maintaining your quality of life.

Anaerobic exercise usually involves resistance. Weight lifting and Pilates are the most popular types of resistive training, and this type of exercise improves your lean body mass, prevents bone loss, strengthens your tendons and ligaments, and increases your basic metabolic rate.

Knowing this, you may wonder why trainers recommend starting with aerobic exercise followed by anaerobic exercise. It's because within thirty minutes after you've gone for a run, the pituitary gland in your brain releases a surge of human growth hormone, peaking for approximately fifteen to twenty minutes. This burst helps to repair damaged muscle mass and increases your muscle mass. It also helps to burn body fat. Your insulin levels also drop, and glucagon—another hormone—rises. Both hormones stimulate another release of growth hormone. Within five minutes of weight lifting, more testosterone and growth hormone are released into your bloodstream, and these can linger for another thirty minutes. So one of the most important anti-aging hormones—the growth hormone—can be manufactured by you, naturally!

### EXERCISE AND NUTRIENTS

You can take certain nutrients and supplements before an exercise routine to boost the release of growth hormone, testosterone, and adenosine triphosphate (ATP) production. This will give you more cellular energy. For many years, I've recommended taking arginine, ornithine, and glutamine—all amino acids (derived from proteins)—to stimulate growth hormone in my patients with low levels of these amino acids. Ideally, you should take these amino acids thirty minutes before beginning exercise. Your testosterone production can be augmented by taking yohimbine extracts. Athletes have used creatine supplements for many years because they act as an energy resource for skeletal muscles. Some scientific studies report that creatine boosts your exercise response and endurance by 5 to 15 per cent.[3,4]

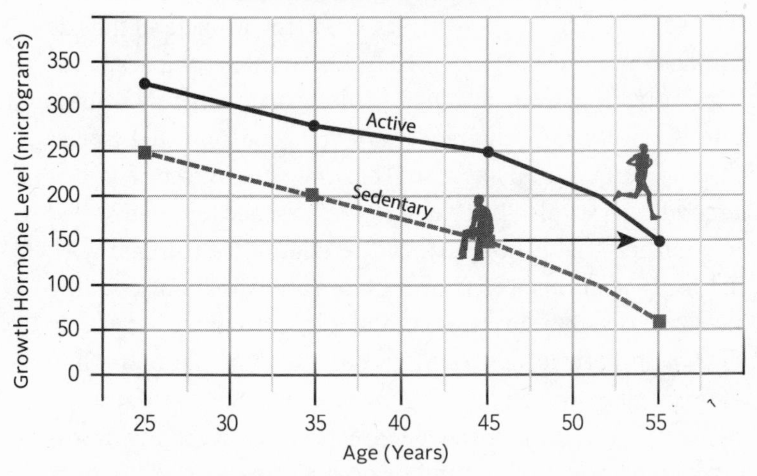


Symbolic representation of what an inactive 45-year-old (dotted line) may have compared to an active 55-year-old (solid line).

### MODERATION APPLIES TO EXERCISE TOO

Is it possible to have too much of a good thing? Should you be concerned about overtraining? Yes. I see this often in my practice.

What drives people to train for one marathon after another, or to ride one hundred miles a day for weeks? Multiple factors, I assure you. It often begins innocently, as a means of exercise or as a personal goal. Some people use it as a means of emotional escape or to fill up “empty” or “lonely” time. Before they know it, the endorphins kick in. These natural painkillers that the body secretes to decrease the muscle pain also create a euphoric high, which becomes addictive. One of my patients told me that most of the people in his bike-riding group are “exercise bulimics”—riding long distances allows them to eat far too much and drink alcohol while staying slim. Although this highlights the psychology of overtraining, there are also many physical downsides to overtraining.

For example, heavy training may actually work against the health of endurance athletes, resulting in skeletal muscle damage. In a South African study, researchers compared athletes suffering from fatigued athlete myopathic syndrome (FAMS) with other athletes. FAMS is a condition that causes chronic fatigue from exercise. The study discovered that athletes with FAMS also had extremely shortened telomeres.[5]

Another study showed that the endurance runners with the highest number of training hours and years of training also have shorter telomeres compared to sedentary people.[6]

What exactly is overtraining? Let's just say that more than sixty minutes of aerobic and/or anaerobic exercise daily can create undesirable health side effects related to aging. Burning more than two thousand calories a week doesn't help you live longer even though it maintains your strength and functionality. Essentially, overtraining stresses your body, which causes the secretion of cortisol. This hormone shuts down a lot of the body's other hormones so that it can switch from growth-and-repair mode to fight-or-flight mode. This means that blood sugar rises (which is bad for anti-aging), and testosterone production drops in favor of cortisol production, creating less muscle mass and increasing the risk of the cortisol-insulin cycle—which impairs healthy metabolism. I discuss cortisol in more detail in Chapter 9.

We also know that exercise produces inflammatory proteins and free radicals. The body can handle the production of some of these—but too much can cause damage.

## Lifestyle Score

The following scoring scale is used to help us assess the lifestyles of some case examples and the impact of lifestyle on telomere length. All lifestyle factors play a role. The research on whether there might be additive or perhaps compounded effects of poor lifestyle on telomere length is not yet available.

We will use this simple scoring card to help get a sense of the quality of lifestyle in each case. The lower the Lifestyle Score, the better the overall lifestyle.

### WHAT IS YOUR LIFESTYLE SCORE?

*Diet*

0: Eats well, does not skip meals, eats balanced meals, eats negligible fast foods
1: Does not skip meals, tends to overeat, eats generally balanced and healthy meals
2: Enjoys a dose of fast foods and junk food but does a reasonable job of eating balanced meals
3: Eats unbalanced meals, tends to eat out, fast foods are often the norm
4: Skips meals, is a poor eater

*Alcohol*

0: No alcohol
1: Social drinker: one to two drinks a week
2: Regular drinker: one to two drinks most nights
3: Heavy drinker: one to two drinks at least per night; three or more on weekends and at events
4: Alcoholic: alcohol dependency, admission to overuse or abuse

*Sleep*

0: Generally good: sleeps seven to eight hours of good, quality sleep
1: Generally satisfactory: some nights six to seven hours' sleep; most nights seven to eight hours' sleep
2: Chronically sleep deprived: sleeps less than six to seven hours
3: Monthly regular episodes of significant sleep deprivation: experiences occasional jet lag of three to five hours of sleep
4: Sleeping difficulties all the time: quality and quantity of sleep are both poor

*Smoking history*

0: No
1: Occasional smoking in teenage years; second-hand smoke
2: Social smoker: a few cigarettes a week—not more than half a pack a week
3: Regular smoker: half pack a week to one pack per day (ppd)
4: Heavy smoker: more than one ppd

*Exercise*

0: Active life: exercises two to three times a week; aerobic and weights
1: Regular activities: plays sports and, occasionally, gym-type activities—running, cycling, and weights
2: No regular sports or aerobic routines: walks only
3: Over-exercising: runs serial marathons and participates in iron-man/iron-woman activities
4: Either completely sedentary or obsessed with exercising more than two to three hours a day

*Score: /20.*

I'd be happy to see my patients have a score of 5 or lower. This means that the most important aspects of your lifestyle—diet, alcohol consumption, sleep, smoking, and exercise habits are in good balance.

On the other end of the spectrum, I'd be very worried if you scored 3 or higher in any one section. There are some lifestyle changes needed immediately, regardless of your telomere length.

In the upcoming cases, a person's telomere length is reported as a percentile relative to a test subject's age group. A percentile score of 75 means that only 25 per cent of this population group (same age and gender) has a better score, and 75 per cent of this same population group has an equal or poorer score than this individual.

CASE STUDY

**Too Much of a Good Thing Is a Bad Thing**

One of my patients has spent much of his adult life cycling—60 miles (97 km) two or three times a week is not unusual. At minimum, he completes two hours of cycling every day as part of his daily routine to and from work (weather permitting). He has never smoked, nor has he been exposed to second-hand smoke; he doesn't drink any significant amount of alcohol. Although his eating habits could be better, he is generally a vegetarian and grazes most days. Yet when I tested his telomeres, I discovered they were shorter than the average.

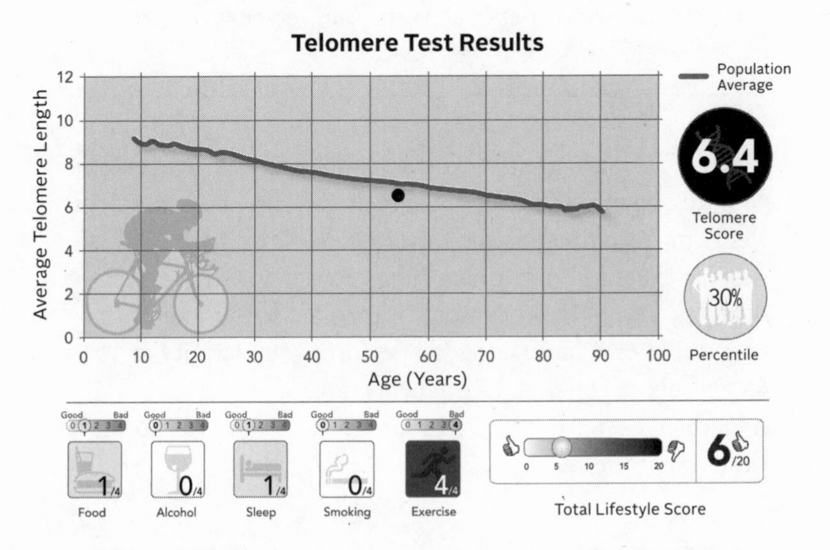


TELOMERE SCORE: 6.4

PERCENTILE relative to patient age and population: 30%

DIET: 1 (Does not skip meals, tends to overeat, eats generally balanced and healthy meals)

ALCOHOL: 0 (No alcohol)

SLEEP: 1 (Generally satisfactory: some nights six to seven hours' sleep; most nights seven to eight hours' sleep)

SMOKING: 0 (No)
EXERCISE: 4 (Obsessed with exercising more than two to three hours a day)
**Score: 6/20**

In my opinion, this client scored in the higher range (remember we're aiming for lower scores as indicators of positive behaviors) because his body has experienced the effects of overtraining. Clearly it is not because he leads an unhealthy lifestyle.

(As with all the examples in this chapter, there are always compounding factors contributing to a telomere length report. Take it for what it is: a snapshot and possibly a report card of lifestyle to date. Bear in mind, genetics and other environmental factors also play a role but are not reviewed here.)

---

We now know that too little exercise is bad for you and your telomeres, and that too much can also be a problem. Here's my recommendation for a reasonable exercise regimen.

- Aerobic exercise: four to five times a week for thirty to sixty minutes. To see improvement in cardiovascular and pulmonary functions, make sure you perform your activity at 55 to 65 per cent of your maximum heart rate.
- Anaerobic exercise: three to four times a week for no more than sixty minutes.

FURTHER INFORMATION

**How to Calculate Your Maximum Heart Rate**

Use the following formulas to calculate your maximum heart rate.
Male: 202-[0.55 × age]
Female: 216-[1.09 × age]
So the maximum heart rate for a 40-year-old woman would be 172.4 (216-[1.09 × 40])

---

Your maximum oxygen consumption (or $VO_2$ max) correlates positively with your telomere length.[7] In other words, the higher your $VO_2$ max, the longer your telomeres. From this, we know that vigorous aerobic exercise is good for you and helps validate its potential anti-aging effects. $VO_2$ max is generally considered to be the best indicator of someone's aerobic fitness. It refers to the maximum amount of oxygen that a person can use during intense exercise. The more oxygen you are able to use during high levels of exercise, the more ATP, or energy output, you can produce. To some extent, your $VO_2$ max is genetically driven, but it can be improved with aerobic exercise.

The ideal method of determining your $VO_2$ max is by measuring your oxygen consumption levels during an intense physical exertion test done by a professional using an oxygen-measurement device. There are also approximation $VO_2$ max calculators available on the Internet. They're not as accurate, of course, but they are "indicative."

What about moderate exercise—can it prematurely shorten your telomeres? The answer is no. While studying older women and men, researchers noted that moderate activity did not have a negative impact on telomeres.[8]

Here is another good reason to get active: exercise increases your antioxidant enzyme activity,[9, 10] and telomerase responds to oxidative stress in a positive way—it may help protect your cells from stress insults[11, 12] and add to the benefits of the oxidative-stress reduction already connected to exercise. However, as I've said previously, higher levels of exercise or overtraining may have the opposite effect.

Another study showed that when it comes to weight lifting, moderate resistance training does not seem to adversely affect your telomere length. Your body appears to be able to repair the muscle tears caused by the workout without causing significant premature shortening of the telomeres.[13]

## THE NET EFFECT OF EXERCISE

Certainly, it's clear that telomere science supports the idea of moderate physical activity to help you achieve Peak Health. Although exercise produces oxidative free radicals, which can damage our cells, the net effect is positive, and your body responds by increasing protective antioxidant enzyme activity. As well, this oxidative stress "turns on" certain hormones and promotes telomerase activity. However, these repair mechanisms can't keep up with excessive exercising.

You should also consider additional alternative exercises as part of your activity regimen, including movement and stretching exercises. Dancing, tai chi, and yoga are all great examples. At the very least, they improve your flexibility, which is critical to preventing injury as you get older, and help promote good spinal alignment and disc health. Movement and stretching exercises also have the important effect of balancing sympathetic and parasympathetic tone in your nervous system, also known as the autonomic nervous system.

This system controls most of the unconscious functions in your body, such as your heart rate, breathing, swallowing, and sexual arousal. These are otherwise known as our primitive functions, and they are essential to our survival. Keeping the autonomic nervous system in good shape and in balance is essential for life.

Another great benefit to getting the right amount of exercise is that you'll likely sleep better too.

Now it's time to discuss breathing. We take our breathing completely for granted. So we tend to neglect it. It's not really considered an exercise, but if you can control your breath, you are essentially in better control of your body. Slow and methodical deep breathing over time improves pulmonary function and oxygen delivery to body tissues and can also help improve waste removal—controlling the body's acidity

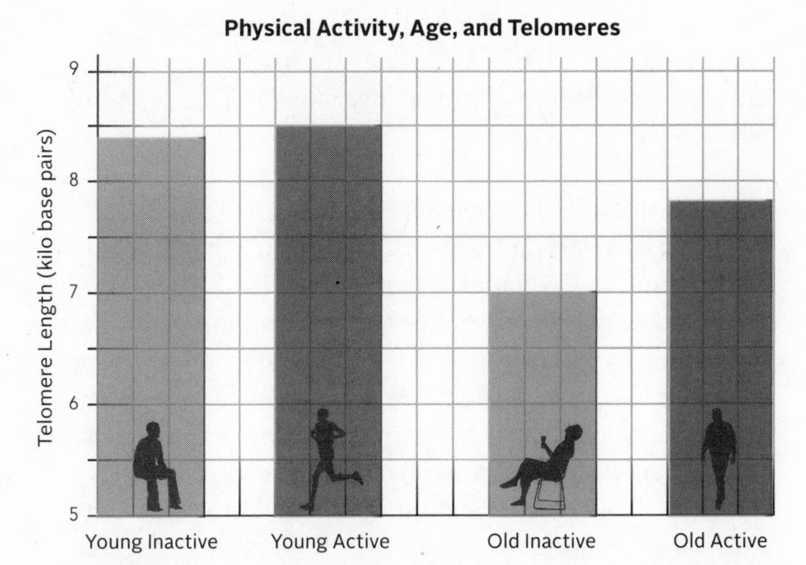


Regardless of age, being active maintains telomere length. It is never too late to start getting active.

(pH levels) to allow cells to function better. Much benefit can be derived from breathing as part of meditation. I touch on that again in Chapter 9.

It's never too late to start exercising. Your telomere length can improve at any age. Whether you walk or run, swim or do tai chi, kickbox or bicycle, it's important to be active at least thirty minutes every day.

## Sleep

Sleep is something we take for granted, and you may not realize how critically important it is for your good health. Although the role of sleep is complex, it's central to your survival and health, in both the short term and long term. Here are just a few benefits of getting enough sleep:

- Sleep is critical for hormonal balance

- Sleep is restorative for wound healing and immune system functioning
- Sleep is needed for growth and rejuvenation in all of our body systems

Sleep affects brain development and brain normalcy later in life and is important for memory and mental processing.

According to the National Institutes of Health, the average adult needs about seven to eight hours' sleep a night. The problem is that many people sleep less than seven hours for long periods of time. This finding is consistent with a survey of Canadians in *Maclean's* magazine.[14] This, in turn, can cause symptoms of sleep deprivation, which can lead to health issues and errors in judgment. Some studies suggest that there are more car accidents and heart attacks the day after we switch to daylight savings time (when we lose one hour of sleep).[15, 16, 17]

A good night's sleep helps us feel ready to tackle another day, and most of us assume we are getting the sleep we need. But a growing number of people are not. A 2011 Canadian study reported that 40 per cent of the two thousand people surveyed experienced insomnia (trouble falling or staying asleep, or early-morning awakening) for a minimum of three nights a week over a month-long period.[18]

Why the growing sleeplessness? Reasons range from stress, sleep dysfunctions (such as sleep apnea and restless legs syndrome), career/job working hours, and attitudinal issues, such as the belief that a person doesn't really need more than five hours a night.

Whatever the reason, many people think that temporary reductions in sleep result only in temporary and "minor" effects, such as a lack of concentration. This may be true in the short term. But there are serious long-term effects we all need to be aware of.

### IMPLICATIONS OF SLEEP DEPRIVATION

We need to understand just how important sleep is to achieving Peak Health and the implications of sleep deprivation. It is during certain stages of sleep that our bodies repair themselves—hence the "refreshed" feeling we have when we wake up. With sleep deprivation, we know that cortisol levels spike.[19] A recent study noted that even a partial acute sleep loss causes a delay in the early-morning recovery of our circadian hormone rhythm, which in turn alters our metabolic and cognitive processes.[20] In other words, we have a harder time thinking straight.

DEFINITION

**Circadian Rhythm** Circadian rhythm is "a daily rhythmic activity cycle, based on 24-hour intervals, that is exhibited by many organisms." *Source: Stedman Medical Heritage Dictionary*

---

In the short term, this alteration leads to a cascade of metabolic reactions, such as higher blood sugar levels and insulin secretion, which both add to an increased potential for glycation. Fred Turek, director of the Center for Sleep and Circadian Biology at Northwestern University, concluded that "short sleepers"—those who slept less than six hours a night—increased their odds for developing obesity and diabetes by 89 and 28 per cent respectively. The difference between having a sleep debt and being fully rested is equivalent to consuming one thousand calories less than usual over three days.[21] Another study researching the metabolic consequences of sleep deprivation identifies short sleep as a cause of obesity.[22]

Clearly, then, we're wrong to believe that transient reductions in sleep result in only temporary, short-term impairment, such as a lack of concentration.

CASE STUDY

### Stress Makes You Fat

A client came to our clinic complaining of a rapid 10 lb (4.5 kg) weight gain over two to three months. He had always been a keen advocate of a healthy lifestyle. He exercised at least three times a week; did weights twice a week; and ate a high-protein, complex-carbohydrate, and low-fat diet. His blood work showed his morning cortisol to be less than 200 (the normal range is between 350 and 500), and his insulin level was high, at 196 (the normal range is between 40 and 80). His blood sugar level and thyroid function were both normal. He denied eating any differently, and we suggested he go on a low-carbohydrate diet for two weeks and then repeat his insulin test. It was still high, but better, at 156.

What happened?

It turned out he had been working on a major real estate deal for the past few months and had trouble sleeping—many nights he only slept two to three hours. Not enough sleep induces cortisol as a stress response, and high cortisol keeps us awake. But as burnout occurs due to chronic sleep deprivation, the cortisol level drops and the body's need to find quick food fuel increases, helping to keep a tired body running. Insulin surges and increases our absorption of any carbohydrate we consume. So, mystery solved. My client gained weight as a stress response.

---

Using a proprietary questionnaire, Symptom Profiler, my studies in 2010 and 2011 that were included in the *Maclean's* magazine "How Healthy Are You?" series, showed that people who slept too much (more than nine hours) or too little (less than five hours) had a higher symptom profile (noted as Q-Gap Score) than those who slept an average of six to eight hours every night.

In the longer term, our response to sleep-loss-induced stress leads to three negative outcomes: impaired immune function, the elevation of pro-inflammatory cytokine production, and a change in the regulation of certain hormones, such as growth hormone and serotonin. All this leads to an increased risk of serious chronic diseases and impaired brain functions.

Scientists from Europe examined sixteen international studies of over one million people who slept less than the recommended six to eight hours of sleep and found that these under-sleepers had a 12 per cent chance of dying earlier than those getting enough sleep.[23]

People say, "Ah, I'll sleep when I'm dead." Well, that might happen sooner than you think if you go without sufficient restorative sleep for a long time.

**Symptom Scores and Sleep**

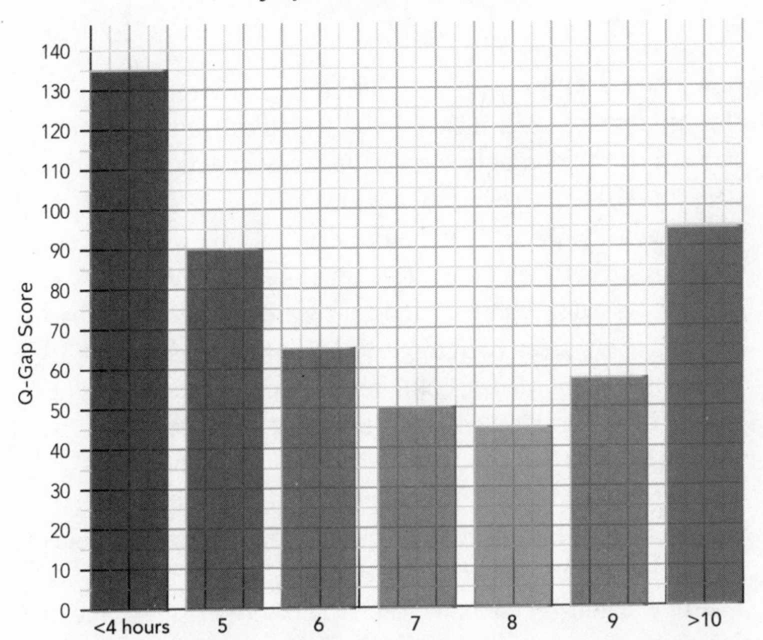


The average number of hours that *Maclean's* readers slept per night, as reported in the Symptom Profiler Survey in *Maclean's*, September 26, 2011.

FURTHER INFORMATION

**Sleep Deprivation Symptoms** You might be sleep deprived if you:

- Need an alarm clock in order to wake up on time
- Rely on the snooze button
- Have a hard time getting out of bed in the morning
- Feel sluggish in the afternoon
- Get sleepy in meetings or lectures, or on the bus or train
- Feel drowsy after heavy meals or when driving
- Need a nap to get through the day
- Fall asleep while watching TV or relaxing in the evening
- Feel the need to sleep in on weekends
- Fall asleep within five minutes of going to bed

Source: the National Institutes of Health website. nhlbi.nih.gov/health/health-topics/topics/sdd/signs.html

---

### SLEEP AND TELOMERES

Now that we know there's a relationship between sleep and chronic disease, let's identify the biological link. Research shows a correlation between the quantity and quality of our sleep and the length of our telomeres.

A study involving ten thousand people showed a direct relationship between telomere length and sleep duration. The men in the study who reported they were getting more than seven hours of sleep each night had telomeres that were longer than those of men sleeping fewer than five hours a night. In fact, the "short sleepers" had telomeres that were 6 per cent shorter on average.[24]

Another study of 245 women showed similar results—women with chronically poor sleep quality displayed telomeres that were shorter.[25]

Are you getting enough sleep? The common-sense way of figuring that out is to think about how you feel during the day. If you're getting enough sleep, you should be energetic

**Sleep and Telomeres**

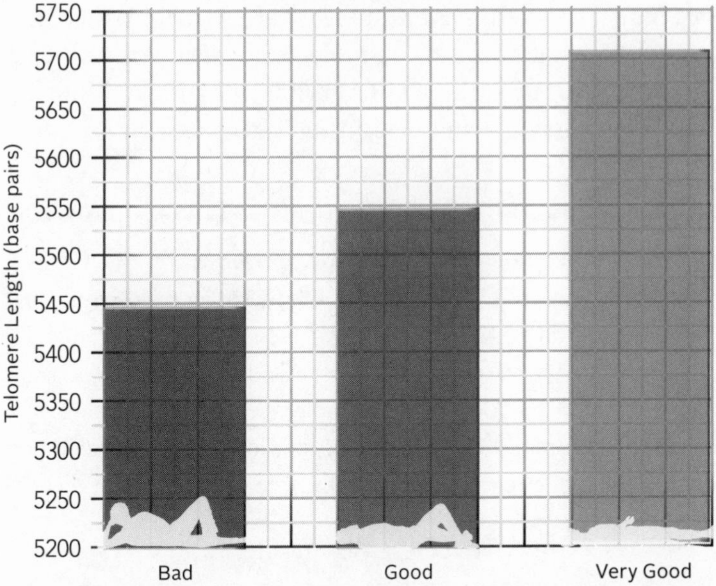

Telomere Length (base pairs)

5750
5700
5650
5600
5550
5500
5450
5400
5350
5300
5250
5200

Bad
Good
Very Good

The more sleep an individual gets per night, the better their telomere length.

and alert throughout your day. And it's not good enough to simply get enough sleep. If you're a poor sleeper, you could be missing out on the sleep cycles that are vital for body and mind. It's not just quantity of sleep—sleep quality matters too to achieve Peak Health.

### CAFFEINE AND ALCOHOL EFFECTS ON SLEEP

Often we forget that caffeine and alcohol are both "drugs" and your brain is affected by them.

Caffeine has a "half life" of about six hours—so that means a cup of coffee at 2 PM is still 50 per cent in your system by 8 PM. Hence it is important to make sure you don't overload or drink anymore after 2 PM if you are having trouble sleeping.

As for alcohol, a new study recently published in the journal *Alcohol* supports the claim that alcohol disrupts sleep. It affects a chemical called adenosine which promotes sleep but induces wakefulness as it wears off. Alcohol stimulates adenosine—so it does get you to sleep, but as it wears off you end up waking up earlier too. Over time, regular drinking can lead to withdrawal symptoms, including insomnia.[26]

CASE STUDY

**Healthy, but Sleep Deprived?**

Some university students skimp on sleep to keep up with their demanding studies. One student we tested eats well generally, doesn't do much exercise, and has never indulged in drugs or alcohol. Hopefully, the demands of schoolwork and staying up late to study will be limited to his twenties, and over time his telomere length will not be "burned" at the same rate.

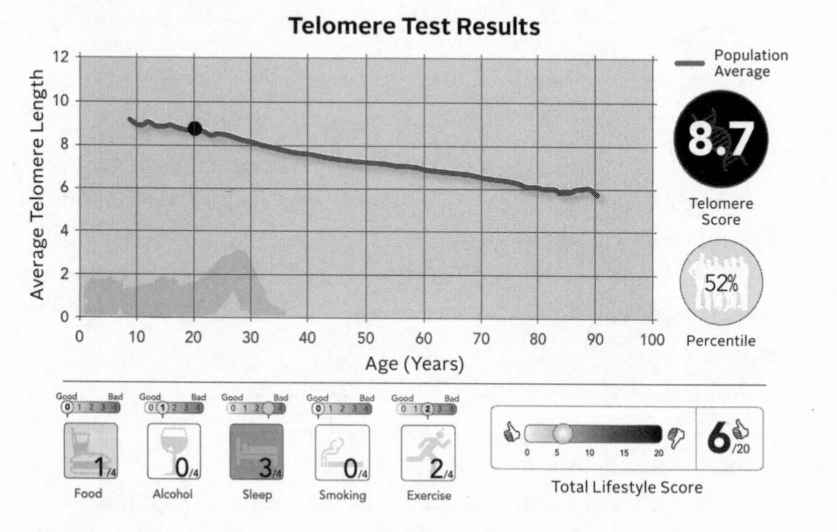


TELOMERE SCORE: 8.7

PERCENTILE relative to patient age and population: 52%

DIET: 1 (Does not skip meals, tends to overeat, eats generally balanced and healthy meals)

ALCOHOL: 0 (No alcohol)
SLEEP: 3 (Sleep deprived)
SMOKING: 0 (No)
EXERCISE: 2 (No regular sports or aerobic routines: walks only)
**Score: 6/20**

My view is that, in this case, even at an early age, sleep deprivation could be a contributing factor to having average—rather than above average—telomere length in healthy young adults.

---

### HOW TO GET ENOUGH QUALITY SLEEP

It helps to get into a proper pre-bedtime mindset. Following these tips can help you create a healthy sleep routine:

- Go to sleep at the same time each day–establish a regular sleep-wake pattern.
- Avoid stimulants (caffeine, energy drinks, amphetamines); these slow the action of sleep-inducing hormones, especially before bedtime.
- Don't nap late in the day–power naps of thirty minutes midday work best.
- Relax by playing soft music or reading before sleeping. Shut down the electronic devices–and that includes the television.
- Create an environment conducive for sleeping–a dark and quiet room is ideal.
- Avoid large or heavy meals close to bedtime–try a light snack containing carbohydrates and protein instead.
- Expose yourself to adequate amounts of sunlight during the day.
- Avoid alcohol before sleeping–it may knock you out, but it will actually keep you from leaving light-sleep stage.

If these tips don't help and you are still unable to sleep properly, consult your doctor, who may prescribe sleep medications. Note that these drugs can be habit-forming, and nearly all are associated with some form of side effect. Long-term use is not recommended.

Natural alternatives to prescription sleep medications are easily available. One of these substances is melatonin, a naturally produced hormone that promotes sleep. Melatonin is secreted by the pineal gland, which is stimulated by exposure to the dark. In turn, melatonin levels peak at night and go down during the day. Because taking melatonin hastens the onset of sleep, it also shows promise for managing jet lag and treating sleep disorders.

*Melatonin: The Basics* Melatonin was not discovered until 1958 by Yale University physician Aaron Lerner. Until his discovery, the pineal gland in the brain, which manufactures and secretes melatonin, was thought to be useless, a leftover remnant of evolution—somewhat like our appendix. We've learned a lot about melatonin ever since, and it turns out to be a very important and powerful hormone for preventing aging and keeping us youthful and healthy.

Melatonin was the first hormone to make big news as an anti-aging medicine hormone since estrogen made it to the mainstream of medical care. (Between 1960 and 1975, the number of estrogen prescriptions in the United States nearly doubled thanks to coverage in the media about the effects of estrogen deficiency and the benefits of estrogen replacement.[27])

The secretion of melatonin slowly declines as we age, and when we reach around 45, it dramatically drops. This triggers responses from other hormones, which begin to also slowly decline, and all of our body system functions begin to slow down. Many anti-aging physicians believe that a supplementation of melatonin can trick the body into thinking it's too early to shut down and that it's time to be young again.

In the early 1990s, Drs. Walter Pierpaoli and William Regelson, co-authors of *The Melatonin Miracle,* tested the function of melatonin. They traded the pineal glands of young mice with older mice. The result? The young mice with the old pineal glands aged rapidly and died 30 per cent sooner than normal, and the old mice with the young pineal glands became more youthful and lived 30 per cent longer than normal.[28]

Every day, melatonin tells our body when to sleep and when to wake up. It acts as a true biological clock, using the changes in light to trigger its release. Light suppresses the release of melatonin. In contrast, as it gets darker, more melatonin is released into the bloodstream, making us feel sleepy.

*Effects of Melatonin* People tend to sleep more in extreme northern and southern climates during the winter months because they don't see as much sunlight. Many people suffer from sleep disorders that are related to lack of sunlight. In contrast, during the summer months, with longer and sunnier days, we all seem to have more energy and feel less sleepy. During the shorter days, our body produces more melatonin earlier in the day than in summer months.

Visually impaired people often suffer from sleep disorders too—they have lost their light-triggering mechanism to regulate their melatonin levels. And because melatonin secretion falls as we get older, it is common for older people to experience sleep disorders. Night-shift workers are also prone to imbalanced melatonin levels.

When we fly from North America to Europe, we find it hard to stay awake even though we have arrived there in the morning. Our calendar clock jumped six hours ahead, but our biological clock didn't follow suit. Similarly, when we go from the East Coast to the West Coast, our calendar clock falls behind by three hours, but our biological clock still wakes us up at

4 AM even though it's still dark outside. This is all due to the melatonin levels in our body. Most sleep experts say that this jet lag phenomenon needs one day to adjust to each one-hour time change.

Teenagers today are showing more evidence of sleep disorders and insufficient sleep. Why? I point the finger at electronic devices; the brightly lit screens affect melatonin secretion. It tricks the brain into thinking it is still daylight. In fact, adults are also experiencing sleep issues for the same reason. We are just as guilty of working late into the night, watching TV, or using iPads and smartphones. Rather than preparing our brains for bed, we are stimulating the wrong hormones, and our melatonin levels are not high enough to give us a restful and long sleep. The answer is to turn off the devices a good one to two hours before going to bed. An alternative is to download a software app that turns down the light of your screens as the calendar clock changes.

Now I know why it was beneficial for me to set all the bedroom lights as bright as possible when I was cramming for exams. Very simply, our bodies lose track of whether it is day or night. When we have no clues from the outside world whether it is night or day, we stay awake longer and sleep fewer hours.

Do you ever notice that if you are constantly lacking sleep, you often fall ill from infections? Being deficient in melatonin can severely weaken your immune system. A study at the San Diego Veterans Affairs Medical Center showed that sleep-deprived men had substantially reduced white blood cell activity.[29] These cells are necessary to protect the body from viruses and bacteria.

As you learned earlier, oxidative stress is not good for our telomeres, and antioxidants are our body's defense against free radicals, the molecules that spread cancerous growth by attacking our cells. Remember also that free radicals target the cell membrane and the cell's DNA, altering the structure of the cell.

Melatonin acts as a powerful antioxidant. It has an advantage over other antioxidants (such as vitamins C and E) in protecting the cells. Not only does it neutralize free radicals out of the cell, but it can also travel into the nucleus of the cell, where our chromosomes are stored. So melatonin is able to directly protect our DNA. In essence, it is an "inside out" antioxidant. It's little wonder melatonin is being studied as a possible cancer-fighting hormone.

*Boosting Melatonin Levels* Melatonin is made from tryptophan. We get tryptophan from the foods we eat. Tryptophan is found in meat and poultry, but the richest sources are sunflower seeds, pumpkin seeds, collards, turnip greens, and especially baked potatoes (with the skins) and bananas. Once tryptophan is consumed, our body converts it into the neurotransmitter called serotonin. At night, serotonin is converted into melatonin.

Melatonin also reacts to temperature. Melatonin levels increase when warm bathwater is heated by as little as 3.6°F (2°C). That's why a hot bath before bedtime can help you fall asleep.

Melatonin can be purchased over the counter as a supplement. It appears to be safe to use in low doses. Its side effects include drowsiness, lower body temperature, and vivid dreams; for some people, it may also affect blood pressure.

You should take melatonin under the supervision of a health professional who is familiar with dosing and the source of the melatonin; melatonin is not a regulated substance, and almost anyone can manufacture it. But some formulations may be of poor quality or impure, increasing your risk of side effects.

Melatonin stimulates the release of several other hormones in the body as well, including growth hormone and the sex hormones. Melatonin tells our bodies to sleep, and it is during REM sleep that growth hormone is released. As a result, some of the benefits attributed to melatonin could be due to the work that other important hormones are doing—another reason why it's important to monitor melatonin dosing.

I can't overemphasize how important sleep is to your health. The next time you pull an all-nighter, remember that there is more at stake than just a good night's sleep. The length of your life is depending on it.

SCIENCE STAT

**Cure for Jet Lag** Flying westbound? Take some melatonin; it will help you sleep longer on the first few nights at your new destination. Nighttime doses of melatonin stimulate sleep. They do not delay or suppress REM sleep, and you won't feel the sluggish effects that other sleep aids can cause. Flying eastbound? Go to bed early instead.

---

**Myths and Facts about Sleep**

MYTH 1: Getting just one hour's less sleep a night won't affect your daytime functioning.

You may not feel noticeably sleepy during the day, but losing even one hour of sleep can affect your ability to think properly and respond quickly.

MYTH 2: Your body adjusts quickly to different sleep schedules.

Most people can reset their biological clock, and even then, by only one to two hours a day at best. Consequently, it can take more than a week to adjust after traveling across several time zones or switching to the night shift.

MYTH 3: Extra sleep at night can cure you of excessive daytime fatigue.

The quantity of your sleep is important, sure. But it's the *quality* of your sleep that you really have to pay attention to. In order to get more mind- and mood-boosting REM sleep, try sleeping an extra thirty minutes to one hour in the morning, when REM sleep stages are longer. Improving your overall sleep will also increase your REM sleep. If you aren't getting enough deep sleep, your body will try to make that up first, at the expense of REM sleep.

MYTH 4: You can make up for lost sleep during the week by sleeping more on the weekends.

Although this sleeping pattern will help relieve part of a sleep debt, it will not completely make up for the lack of sleep. What's more, sleeping later on the weekends can affect your sleep-wake cycle so that it is much harder to go to sleep at the right time on Sunday nights and get up early on Monday mornings.

FURTHER INFORMATION

**Tips for Getting and Staying Out of Sleep Debt**

- Aim for at least seven and a half hours of sleep every night. Make sure you don't fall further into sleep debt by blocking off enough time for sleep each night. Consistency is the key.
- Settle short-term sleep debt with an extra hour or two a night. If you lost ten hours of sleep in a week, pay the debt back in nightly one- or two-hour installments.

*From the U.S. Department of Health and Human Services,* Your Guide to Healthy Sleep, *The National Institutes of Health, National Heart, Lung, and Blood Institute.*

SCIENCE STAT

**Sleep Apnea and Telomere Length** We know that sleep apnea is a sleep disorder and is not historically linked to lifestyle. But we also know that people who are obese are more likely to develop sleep apnea. It is characterized by abnormal pauses in breathing; these breaks can be as short as ten seconds or they can last minutes, and they can happen as many as thirty times or more an hour. This abnormal breathing behavior leads to low oxygenation.

A group of French scientists showed that telomere length was much shorter in patients with OSAS (obstructive sleep apnea syndrome).[30] If left untreated, conditions that disrupt sleep, such as sleep apnea, can cause heart disease, pulmonary hypertension, and high blood pressure, and this has also been linked to type 2 diabetes.

---

## Alcohol and Smoking

You don't need to see your telomeres under a microscope to know that smoking and drinking don't mix with good health. Using the two most common legalized recreational drugs in the world is one surefire way to burn up your telomeres.

Let's start with alcohol. We've come a long way with curbing drinking and driving, and the combination of drinking and health deserves just as much attention. Studies show that your telomere length can decrease in relation to the number of alcoholic drinks consumed in a day. People who drank more than four drinks a day showed telomeres that were half the length of those enjoying less than one drink a day.[31]

Heavy alcohol consumption has also been linked to increased oxidation stress and inflammation (and we know how those two things can speed up telomere shortening) as well as to your cancer risk.

**Alcohol and Telomeres**

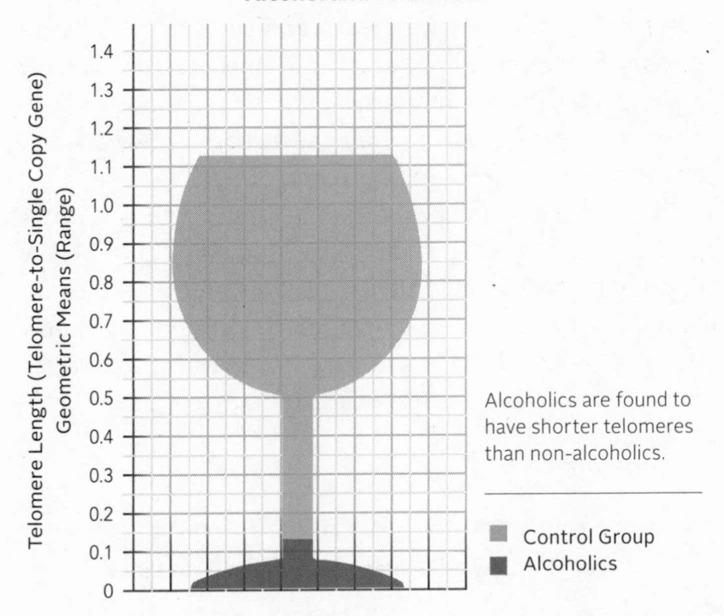


Personally, I don't think that even one drink a day is safe, at least from a cancer perspective. Although there are data to suggest that a drink or two a day is cardio-protective, the opposite is true for cancer. In fact, your risk of mouth and pharynx cancer goes up by 17 per cent if you consume just one drink a day; breast cancer risk in women goes up by 5 per cent. Worldwide, every year, over two million deaths are linked to alcohol, according to a report in the *Annals of Oncology*—and the same report attributes 3.6 per cent of all cancers to drinking alcohol.[32] Dr. Thomas Sellers at the Mayo Clinic found that women who are daily drinkers and had close relatives with breast cancer doubled their risk of developing breast cancer compared to women who also had close relatives with breast cancer but never consumed alcohol.[33] So, is that drink really worth it?

To bring the message home, a recent Harvard study showed a dose-dependent risk of the number of alcoholic drinks and the risk of developing breast cancer. Women who drank three to six drinks a week increased their breast cancer risk by 15 per cent, and those who drank an average of two drinks a day increased their risk by 51 per cent.[34]

To be frank, I'm shocked by the "safe level" guidelines published by the Canadian Centre on Substance Abuse, which claim that women should consume no more than two drinks a day, and men no more than three. From my experience, the low-risk level is *occasionally* one drink a day. Period. The only way to avoid disease risk attributed to alcohol is by abstaining.

SCIENCE STAT

**Canada's Low-Risk Alcohol Drinking Guidelines**

**"A Drink" Means:**

1.5 oz/45 mL serving of hard alcohol, such as rye, gin, rum, etc. (40% alcohol content)

5 oz/142 mL glass of wine (12% alcohol content)

12 oz/341 mL beer, cider, or cooler (5% alcohol content)

And if you light up a cigarette or cigar with your drink, you're just asking for trouble. Watch as that cigarette burns down. It's the perfect metaphor for telomere shortening—shown at high speed. Simply put, tobacco smoking shortens telomeres faster than not smoking, as shown by Ana Valdes et al. in a 2005 study.[35]

For those who say they are only a social smoker or just have cigars when they're golfing, the data suggest that any smoker, regardless of whether they drink alcohol or not, has shorter telomeres. Like alcohol, there is no "safe dose" for tobacco smoking when it relates to telomere health, but there is a dose-dependent effect. The more cigarettes you smoke in your life, the shorter your telomeres will become.

There is a lot of information to support this claim. First, we know smoking can cause inflammation throughout the body; second, it's well known that it also causes oxidative stress.[36] These two processes have been well discussed in earlier chapters, and we know that inflammation and oxidation are two of the four big health enemies in telomere health, along with glycation and hormone imbalance.

### Maintaining a Peak Health Lifestyle

Our lifestyle is so important to our health—and how robust our lifelines remain. But beyond the physical realm of diet and exercise, there is another sphere of life that takes a huge toll on us that we underestimate. The next chapter goes beyond the physical to the effects that our mental health has on us. In order to work toward our Peak Health, we need to take a close look at how stress is affecting our bodies and our lives.

CASE STUDY

#### Bad Habits

Bad habits can catch up to you; they might catch up to your telomeres first. One of my patients admits that he has certainly lived the "bad" lifestyle, and he attributes it to an unhappy marriage. He drank "a lot of Scotch" and "tons of wine," plus he smoked three packs a day. The good news is, today he drinks occasionally on a social basis, he's stopped smoking, and his diet is full of nutritious foods. His telomere report is a representation of his life to date.

Hopefully, moving forward with his new healthy lifestyle, we can preserve his remaining telomeres.

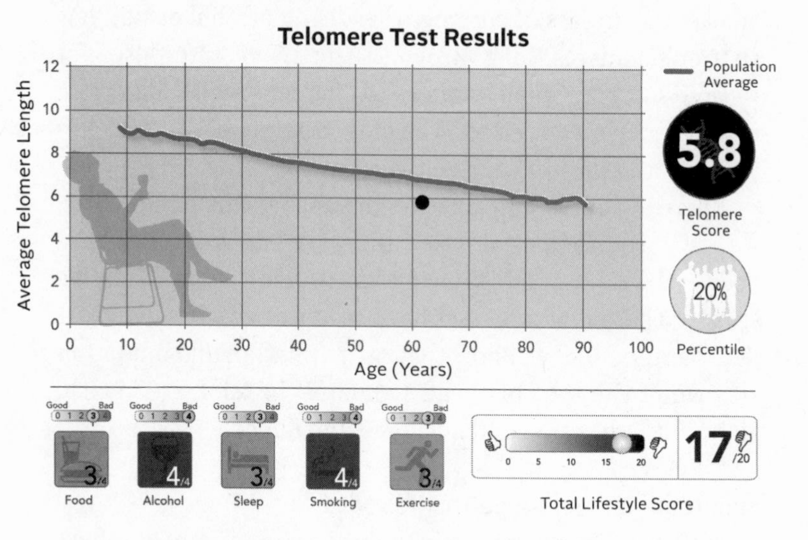


TELOMERE SCORE: 5.8

PERCENTILE relative to patient age and population: 20%

DIET: 3 (Eats unbalanced meals, tends to eat out, fast foods are often the norm)

ALCOHOL: 4 (Alcoholic: alcohol dependency, admission to overuse or abuse)

SLEEP: 3 (Monthly regular episodes of significant sleep deprivation)

SMOKING: 4 (Heavy smoker history; more than 1 ppd)

EXERCISE: 3 (Over-exercising: runs serial marathons and participates in iron-man/iron-woman-type activities)

**Score: 17/20**

This case study confirms that bad things happen to our telomeres when we abuse our bodies with bad lifestyle habits.

# 9

# Mental Health

*We now know that stress seeps into the cell and changes hundreds of biological processes, including the rate of aging.... Telomeres are the clock or pacemaker of the cell's life.... It appears our cells are listening to our suffering and eavesdropping in on our thoughts. The implications [of this mind-cell relationship] are vast.* DR. ELISSA EPEL, PhD, Department of Psychiatry, University of California, San Francisco[1]

## Mental Health Affects Physical Health: A Proven Relationship

Finally, we have proof of what medicine has long suspected: our mental health can drive our physical health. Earlier, I explored how telomeres shorten as we age and how they can predict risk factors for disease. In this chapter, I focus on the effects of our mental health on telomere biology, and how we can reverse the harmful effect of stress on our lifelines.

Stress leads to shorter telomeres. We finally have biological proof that chronic physical diseases are intertwined with a lack of mental well-being. On top of that, the reality of the global economy is that we are surrounded by more stress than ever. So, our mission has to be: we all have to focus on reducing our negative stress. Otherwise, we'll be getting sicker rather than better, and our health care systems, already heaving from overuse and spiraling costs, will collapse. So it makes sense to focus on managing stress so that we can live well—and live long.

Many small studies have found that people who have been suffering from depression for long periods of time have shorter telomeres. In fact, the rate appears to be in line with the total number of days a person is depressed over their lifetime. Oxidative stress and inflammation seem to be related to this process.[2] In other words, our physical maladies are related in some way to our mental ones.

A groundbreaking study in 2004 highlights the link between stress and the rate of telomere shortening. The researchers looked at the telomere lengths of mothers who were responsible for the care of their chronically ill children.

The longer a mother spent time as the main caregiver of a child affected by a serious disorder, the shorter her telomeres were compared to mothers caring for children who were generally healthy. They also found that mothers who felt a sense of control over their lives had longer telomeres than mothers who felt their lives were more stressed. Quantified, the "most stressed" mothers exhibited telomere shortening equal to at least ten years of aging.[3] This means that this stress could shorten their lives by ten years.

Many researchers have strengthened the cause-and-effect connection between chronic stress and shorter telomere length since that study was published. In fact, they have found that chronic stress may indeed *cause* our telomeres to shorten (not just that there is a *link*). A 2011 study showed that this causal effect appears to start at a very early age—in childhood, and perhaps even before a child is born. It found that young healthy adults of mothers who experienced relatively stress-free pregnancies display longer telomeres than their peers whose mothers experienced severe stress while pregnant—such as the stress that happens after the death of a close family member.[4]

CASE STUDY

### Shorter Telomeres Can Be a Product of Early Stress

One of my patients was adopted as a young girl. Her birth mother had been a teenager—maybe 15 or 16—when she gave birth. It's likely that her birth mother experienced a lot of stress during her pregnancy and during the adoption process. Could this have contributed to my patient's below-average telomere length? During her childhood, she didn't experience any significant stress in her adopted home. She recalls it as a loving and nurturing one. As well, she has been generally healthy throughout her life and maintains a very conscious, healthy lifestyle.

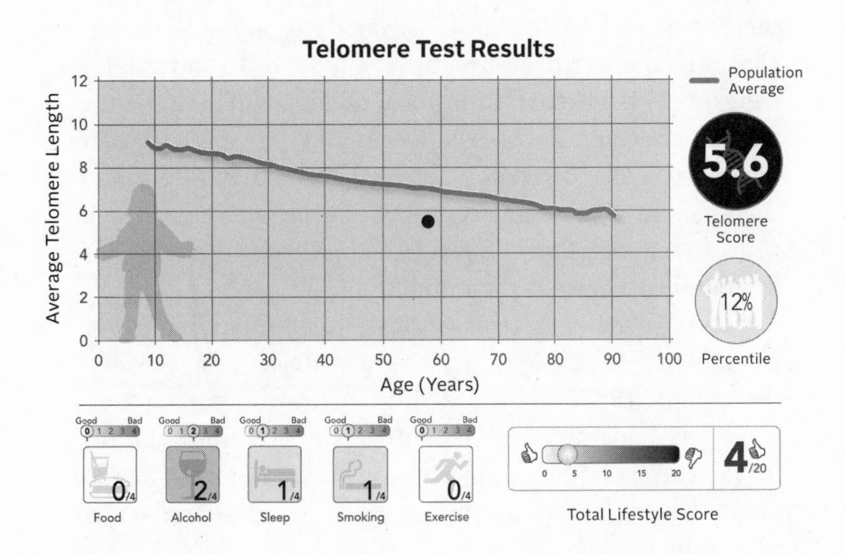

TELOMERE SCORE: 5.6
PERCENTILE relative to patient age and population: 12%
DIET: 0 (Eats well, does not skip meals, eats balanced meals, eats negligible fast foods)
ALCOHOL: 2 (Regular drinker: one to two drinks most nights)
SLEEP: 1 (Generally satisfactory: some nights six to seven hours' sleep; most nights seven to eight hours' sleep)
SMOKING: 1 (Occasional smoking in teenage years; second-hand smoke)
EXERCISE: 0 (Active life: exercises two to three times a week; aerobic and weights)
**Score: 4/20**

This case questions the effects of early "womb" stressors and their impact on telomere length.

---

We also know that the stress experienced in your early years can affect your life later on. In a study that looked at over four thousand middle-aged women, researchers found that women who experienced adversity as a child (divorce, physical abuse, or drug use, for example) had shorter telomeres, and that there is an inverse relationship between the number of their adverse experiences and the length of their telomeres. Violence takes a toll. And the study also found that spending time in an orphanage has the same effect. Even going to school for fewer years in a child's early life is linked to shorter telomeres in their middle and old age.[5]

The message is clear. If you can't reduce chronic stress, whether it's from physical or sexual abuse, financial distress, domestic violence, bullying, victimization, or physical or emotional neglect, you will experience illness. I'll say it again: stress injures. In fact, stress kills. Given this, we can view telomere length as a psychological biological marker, or psycho-biomarker.

DEFINITION

**Stress** Lazarus and Folkman's Stress and Coping Theory defines stress in the following way: a feeling of stress occurs in "situations where a goal that matters to the person is at stake, and the demands of the situation outweigh the person's resources for coping with it."[6]

Positive stress, or what we might call a challenge, occurs when a person can find sufficient or nearly sufficient resources to meet a situational demand. In contrast, negative stress occurs when a person feels he or she does not have sufficient resources to meet a situational demand.

An example of positive stress is expecting a newborn baby in a welcoming home; a negative stress might occur when someone is ill and can't afford the medications necessary to treat their illness.

---

## Hormones and Stress: Cortisol

We've all heard the phrase "Stress kills." Now we know *how* it kills. When we are stressed, our body responds with biochemicals—notably one hormone, cortisol. This is our fight-or-flight-response hormone. It seems that this biochemical activates telomerase in response to acute stress. However, when the stress becomes chronic, the response is blunted or dampened—and the shortening of our telomeres begins. Earlier, I explained the roles that serotonin and melatonin hormones play in our sense of happiness and well-being. We know that happier people live longer and that they have longer telomeres than unhappy people.

I should also add that hormones—insulin and growth hormone—also play an important role in how we respond to stress, whether it is physical, emotional, or both. Recall that they build and repair our bodies in response to our stressors. Although these hormones have other functions beyond stress response, I'm focusing on them as a group because we know stress plays a big role in the premature shortening of telomeres.

Cortisol hormone, along with DHEA, is produced in the adrenal gland. This stress-response hormone is responsible for getting our bodies ready for a fight. Our heart rate, blood pressure, and body temperature all rise to increase our metabolism in response to stress. Unfortunately, these days (and all too often), this stress response is caused by an emotional rather than a physical sense of danger. Nevertheless, the response is the same.

Short-term cortisol is important for preserving our well-being. Our immune response goes into high alert, preventing us from getting infections and becoming ill. Our DNA and tissue repair response increases. We are stimulated to find food—especially carbohydrates—and we store it as fat so that we are less likely to starve. This is an example of short-term stress.

Long-term stress, however, can reduce our cortisol response—often known as adrenal fatigue or burnout. This leads to a slower immune response, which can complicate how we prevent and repair DNA damage, which, in turn, increases our risk of infection and disease. As well, with a slower metabolic rate, we have less energy, feel more tired, and burn less fat.

If you have a smartphone, you know what happens if you don't keep it charged. You can think of your cortisol levels in a similar way. A fully charged smartphone battery is usually good for a full day of activities. However, if we do an unusually high amount of talking or online surfing, the battery runs out quickly and doesn't get us through the entire day.

Our bodies need recharging too. If we get a full night's sleep, we replenish our cortisol levels and are powered up for the day. If we don't sleep well or long enough, we begin the day exhausted and drag ourselves through our day, feeling completely burned out by night time.

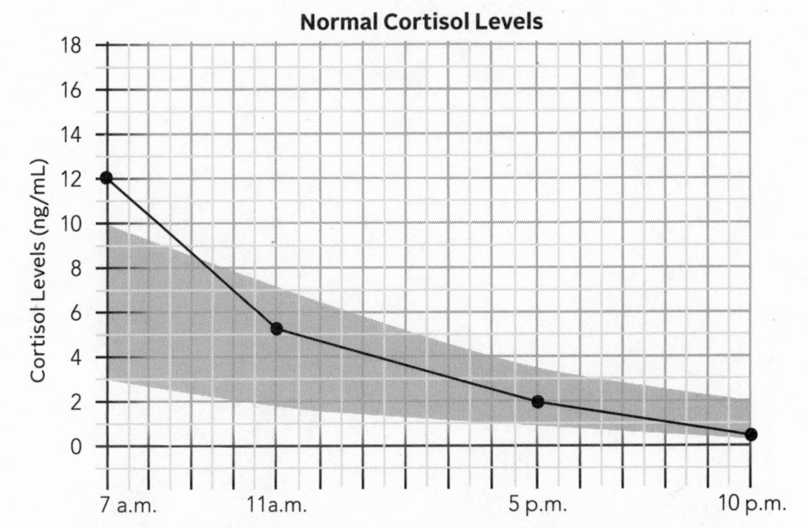


An example of a healthy cortisol curve within the shadowed normal range. As the day progresses, cortisol levels gradually decline.

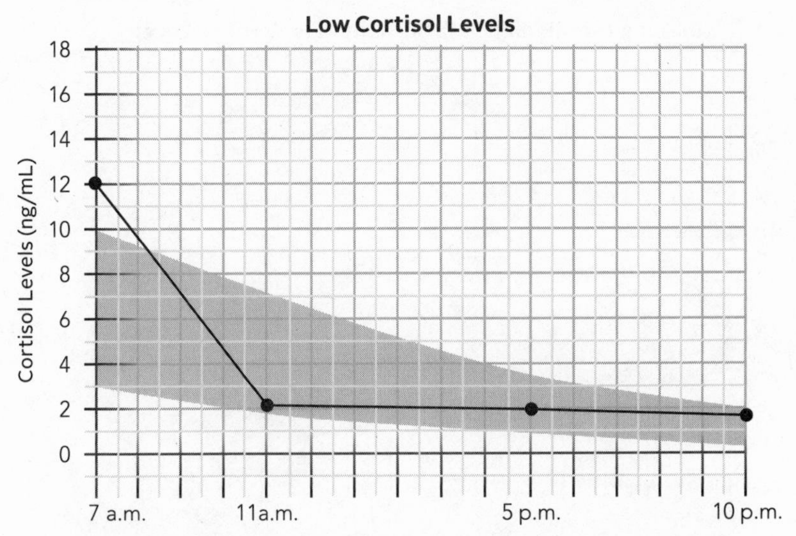


An example of early burnout. This individual is fully charged up in the morning but does not have the "stamina" to maintain a gradual healthy decline of cortisol level.

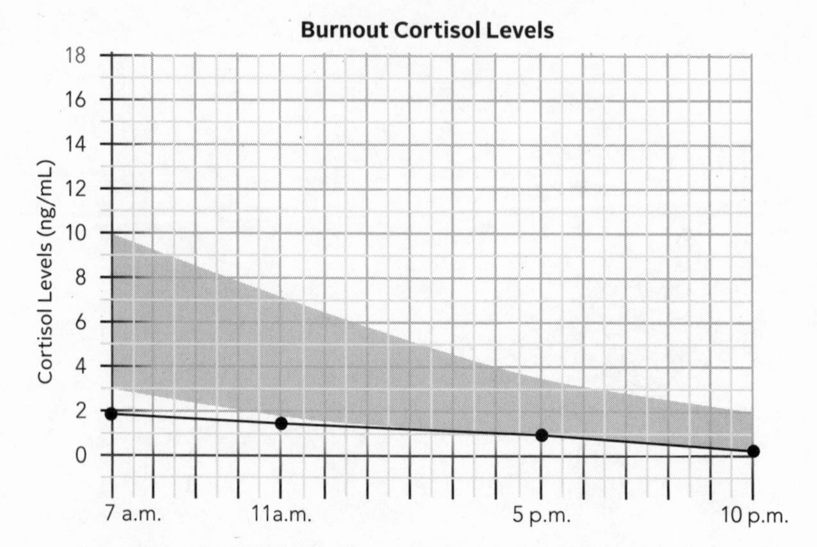


An example of exhaustion or "burnout." There is no recovery of cortisol the night prior and the morning begins with a deficit level of cortisol.

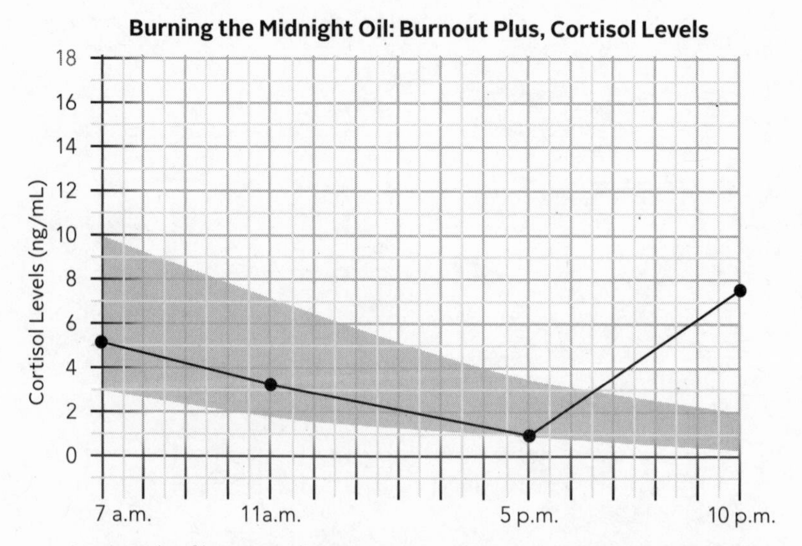


An example of burnout, where someone pushes past the exhaustion level. This individual gave a speech around 7 pm after a long workday. When it's time for bed, cortisol levels are too high, and in this "wired state," sleeping well is difficult.

Unfortunately, there is no hormone-replacement therapy for modulating cortisol response. When we are stressed or burned out, we need to moderate and change how we live. Your most important behaviors to maintain healthy cortisol levels are to exercise regularly, to get good-quality sleep, and to avoid stimulants. This means that we must stay away from substances such as alcohol, caffeine, and sugar; they create a yo-yo effect of higher "highs," which lead to lower "lows." Other effective methods for maintaining healthy cortisol levels are daily meditation and yoga, as well as getting a weekly massage.

In addition, with the guidance of a health practitioner, there are some supplements that can help stabilize abnormal cortisol levels and, in essence, support your adrenal gland function. They include glandular therapies, hormone modulators, and B vitamins.

### Alleviating Stress

Negative thoughts, especially those that spark a sense of threat or cause obsessive thinking, can lead to an urge to react to a situation. Studies show that reducing this sense of arousal or panic can come about with mindfulness and meditation techniques, such as deep breathing and positive imaging.

### What Is Mindfulness?

The way Western society views mindfulness is different from traditional Buddhist teachings. Buddhist traditions define mindfulness more as an ability to maintain attention on certain meditative objects, such as a word, a physical object, or a mental picture.[7] There's also an emphasis on concentrating on beneficial thoughts. Western mindfulness is often more psychological and includes carefully observing and labeling internal and external experiences in nonreactive and nonjudgmental ways.[8] One reality remains: no matter how you do it, being mindful can reduce stress.

SCIENCE STAT

**Serotonin** Serotonin is manufactured in the brain. I call serotonin our happiness hormone because it regulates our mood and sexual desire. As well, serotonin affects our appetite, sleep, memory, and learning. People who are deficient in this hormone often suffer from symptoms of depression, obsessive-compulsive disorder, and anxiety.

As it gets dark each day and evening arrives, our serotonin hormone level begins to rise. The function of serotonin is to balance two neurotransmitters, dopamine and adrenaline, allowing our bodies to rest. If we were bombarded with "wakening" hormones all day, we would quickly become exhausted.

The exact level of serotonin in the brain required for positive mental health remains a mystery. Another unanswered question: does a sudden drop in serotonin cause depression?

Once it was discovered that there is a connection between serotonin and depression, doctors were able to treat this illness, in part, as a hormone disorder. If you've been feeling down, take a look at the mini depression screening test on the Internet. It's just a few questions, but they can reveal a lot.

We do know that what we eat can influence our levels of serotonin. In particular, tryptophan, an amino acid that is used in making serotonin, is found in protein. Turkey, in particular, is known for its high level of tryptophan, as are nuts, milk, and other dairy products. Getting enough vitamin $B_6$ can affect the rate of conversion of tryptophan to serotonin.

Both tryptophan and serotonin (as a precursor called 5-HTP) are available through health professionals as a supplement. People who experience mild symptoms of depression or seasonal affective disorder often find them helpful in reducing the sensation of being "low" or "feeling down."

FURTHER INFORMATION

Jon Kabat-Zinn is Professor of Medicine Emeritus and the founding director of the Stress Reduction Clinic and the Center for Mindfulness in Medicine, Health Care, and Society at the University of Massachusetts Medical School. He teaches a program called Mindfulness-Based Stress Reduction (MBSR) and has written a number of books on the topic of mindfulness. One of the most popular is *Wherever You Go, There You Are: Mindfulness Meditation in Everyday Life.*

Can we counteract the effects of stress on telomeres? The world of medicine believes that telomeres are stable structures that shorten slowly over many years. Some scientists had assumed that our telomere length does not change over a period of months. However, some small studies have found the opposite to be true. Telomerase activity can be increased and can respond to certain lifestyle and mindset changes quite quickly.

In an uncontrolled study, celebrated physician Dr. Dean Ornish oversaw an intensive lifestyle modification program for men with prostate cancer. Those who followed a plant-based diet (high in fruits, vegetables, and unrefined grains, and low in fat and refined carbohydrates), increased their physical activity, and used stress-reduction techniques (such as yoga, meditation, and social support) saw their telomerase activity increase by 30 per cent over three months.[9]

In another study, overweight women followed a program that involved stress reduction and mindful eating. The study found that telomerase activity increased by 18 per cent in a treatment group compared to the control group.[10] What's more, those women who showed the largest decreases in psychological distress, cortisol, and glucose also showed the

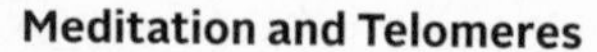


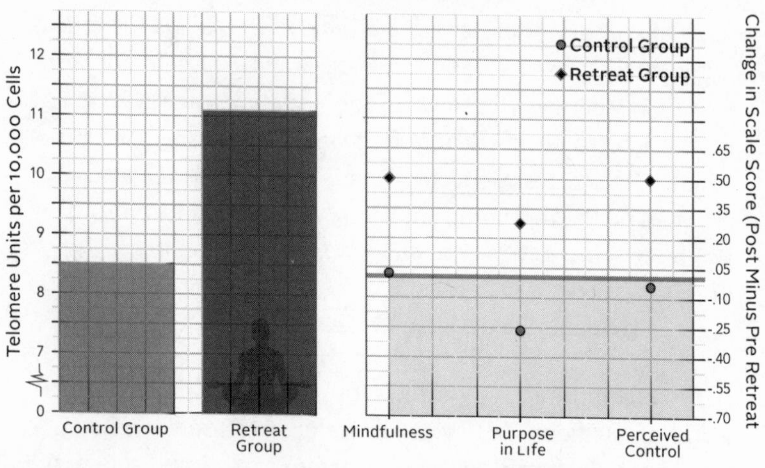


This study clearly showed that those who practice mindfulness, have a life purpose, and appear in control of their lives have longer telomeres.

greatest increases in telomerase activity. (Recall that lower cortisol levels lead to lower insulin, which stabilizes blood glucose levels. When this occurs, your aging processes are reduced.) I explain later how meditation could boost telomerase activity.

In a study of dementia caregivers (men and women who care for patients with dementia), a group practicing meditation showed more telomerase activity—43 per cent more—compared to a relaxation control group.[11]

A more definitive study investigated the effects of a three-month meditation retreat on telomerase activity and two major stress contributors: perceived control (associated with decreased stress) and neuroticism (associated with increased subjective distress). Thirty retreat subjects meditated for three months—six hours each day. The researchers looked at mindfulness as one quality cultivated by meditative practice and at another quality developed during meditative practice.

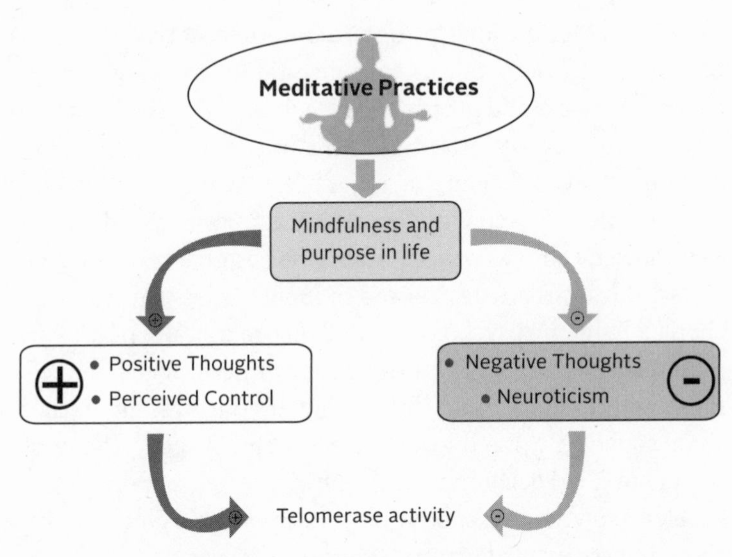


A flow chart of what meditative practices can achieve to activate or deactivate telomerase to preserve and lengthen our telomeres.

This quality hasn't been studied in depth—a shift in intentions and priorities, away from boosting your "hedonic" pleasure (superficial well-being), to making a deeper "purpose in life" more clear.

The researchers found that telomerase activity rose significantly in the meditation group compared to the control group. The retreat group also displayed increases in perceived control. Mindfulness and purpose in life were also greater.[12] This study was one of the first to link meditation and positive psychological change with telomerase activity. A more recent study of distressed breast cancer patients showed that psychological interventions involving mindfulness meditation and gentle Hatha yoga helped to maintain telomere length compared to the control group who received the usual stress management care in a group support setting. Their telomere lengths were found to have been decreased.[13]

**How Does Meditation Prompt Telomerase Activity?**

Buddhist traditions tell us that meditation can reduce psychological stress and promote well-being. In one study, meditation improved the attention skills of the participants. The authors believed that enhancing "task-relevant attention and focus rather than straying toward uncertainties and difficulties allowed those who meditated to be better able to perceive stressful life circumstances and thoughts as less threatening, thereby reducing psychological and physiological stress."

In other words, you can train your mind to focus on feelings and thoughts, and on staying in the present, rather than constantly having negative thoughts about the past or future.[14, 15] Focusing on the "now" creates a more active problem-solving perspective on your life than a more passive thought pattern of alarm and self-doubt does.

It's clear that how you perceive and cope with stress is critical to preserving your telomere length. Lazarus and Folkman's Stress and Coping Theory defines coping as "constantly changing (moment to moment) cognitive and behavioral efforts to manage the demands of a stressful situation."

If you evaluate a situation and believe there is a good chance you can control it, this will induce you to find a more active problem-focused behavioral response. For example, if you've fallen behind in your mortgage payments, rather than panic, you find a way to resolve the situation, such as getting a second job to support those payments. Or you're told you have cancer and that it is treatable. Rather than worry that you'll die, you focus on the surgery and recovery at hand. However, if you assess a situation and decide that there is nothing you can do to alter the course and you just accept or give in to it, you will begin a more "cognitive," or mind-based, strategy that allows you to emotionally accept the situation. This impacts the health of your telomeres because your emotional response to stress is connected to telomere shortening, as the recent findings about the breast cancer survivors attested. Using the same example above, you can resolve to sell the house and move into a home with a more manageable mortgage.

The effects of stress on telomere shortening may build up over time rather than be caused by a single event, although how severe the stressor is and the perceived threat also play roles.

CASE STUDY

**Young Patient with Short Telomeres. Why?**

This patient in her early forties was surprised by her shortened telomeres. She has always been active, she's never smoked, and she drank only socially. Her lifestyle was optimal. She eats well and augments her diet with supplements. So why were her telomeres shorter than those of other women the same age?

It's possible that two dramatic emotional traumas affected her emotional health. First, her mother died when she was a teenager; second, her best friend died in an accident during university.

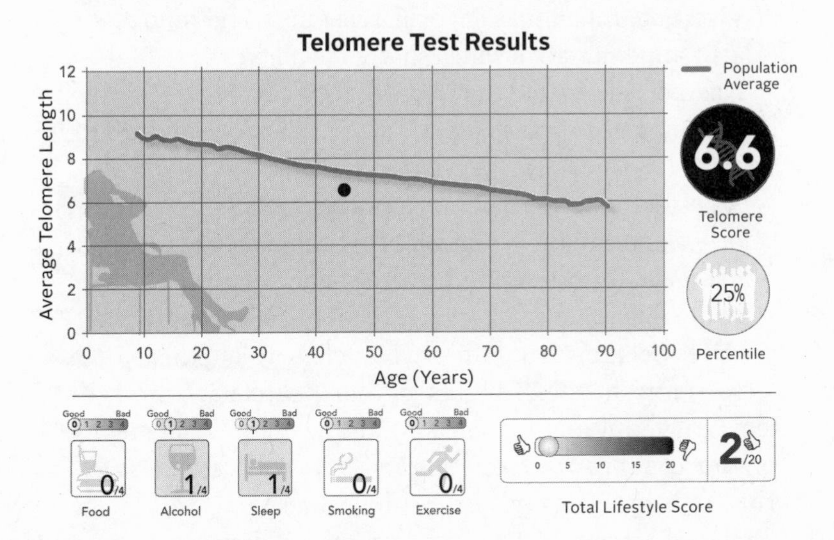

TELOMERE SCORE: 6.6
PERCENTILE relative to patient age and population: 25%
DIET: 0 (Eats well, does not skip meals, eats balanced meals, eats negligible fast foods)
ALCOHOL: 1 (Social drinker: one to two drinks a week)
SLEEP: 1 (Generally satisfactory: some nights six to seven hours' sleep; most nights seven to eight hours' sleep)
SMOKING: 0 (No)
EXERCISE: 0 (Active life: exercises two to three times a week; aerobic and weights)
**Score: 2/20**

For me, this case demonstrates how life's emotional insults can affect our mental health and, in turn, our physical biomarkers.

---

## Positive Thinking Pays Off

We now know that stress affects telomere shortening in a number of ways. As we learned earlier, stress affects our immune and hormonal systems. This, in turn, has an impact on glycation, inflammation, oxidative stress, and mitochondrial dysfunction, all of which shorten telomeres.

Another landmark study by Dr. Elissa Epel and her colleagues provided a physiological theory on how we respond to stress.[16]

A positive emotional state gives a person a sense of well-being, balance, and self-control.[17] This stimulates a hormonal response of testosterone, dehydroepiandrosterone (DHEA), and growth hormone secretion (I've discussed at length their positive effects on maintaining your telomere length).

In contrast, stress messes up our neuroendocrine system, which can result in a change in the diurnal (day-night) rhythm of our cortisol secretion and elevate its baseline level. This, in turn, increases our insulin production.[18, 19, 20] These combined actions lead to glycation and oxidative stress. By

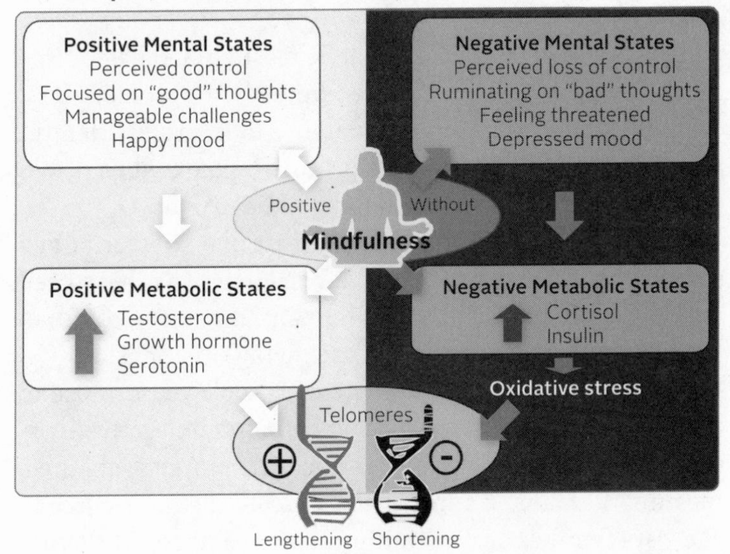


now, you know that both of these are known biomarkers of aging and telomere shortening.

Knowing all this, how is it possible to accurately determine, or "read," when a situation will be critically stressful enough to affect the biology of your telomeres? When you read a stressful situation appropriately and use the proper coping strategy, the hormones that optimize your telomere health are secreted. But if the opposite happens (for example, you try to control an uncontrollable situation, like trying to stop your daughter from getting married to a man you don't like), your stress will not subside and destructive hormones will be secreted, damaging your telomeres. In this way, you can see how *perceived control* is a key indicator of stress resilience.

There are now many studies that support this biological theory. Objective stress happens when you feel trapped in a situation. This can occur, for example, when a person must provide caregiving for many years or when there is general perceived chronic stress—both will result in shortened telomeres.

It's hard to predict who might be most vulnerable to telomere shortening when they're exposed to chronic stress. What we can do is improve our ability to interpret stressful situations properly. Essentially, it's good to know when to fold your cards, when to ask for help, and when to fight.

Most of us are dealing with at least one stressor at any given time, if not many different ones—ones that have been with us for years. Such as? Well, financial anxiety, dysfunctional relationships, or difficult caregiving situations.

So let's consider the healthy coping tools you can use to stay positive. You can start by channeling your negative emotions (such as fear, anxiety, and panic) into goal-oriented, meaningful plans. It's interesting to note the differences in the ways families deal with illness. Some embrace the situation and find higher purpose and reasons for why the illness has struck them, while others can't move forward even day to day. Indeed, people who lead a "faith-based" life often live longer than those who don't.[21, 22] Perhaps it has something to do with an existing coping strategy—a faith that your life is manageable and in the hands of a higher power.

I'm not advocating that you should become religious. Rather, I'd suggest that we must all develop and practice a value system that fosters higher purpose, setting our goals beyond those that our materialistic society says are good for us—so that when adversity hits, we can attribute a "meaning" to it and discover a way to reason with the situation, and to control it.

This brings me back to meditation and other contemplative practices. They all promote a sense of direction and purpose for our lives, and our priorities shift away from self-gratification and hedonic pleasures to more genuine contentment and to a stronger sense that we are contributing to our community. Many studies, including a growing number of "happiness" studies, support the idea that when we see life as meaningful, we automatically use more flexible coping strategies and have greater resilience around stress.[23]

I was recently struck by a lecture I attended at the Rotman School of Management in Toronto, Canada. The venerable Tenzin Priyadarshi is the executive director of the Dalai Lama Center for Ethics and Transformative Values at the Massachusetts Institute of Technology (MIT) in Boston and he is a Tibetan Buddhist monk. He teaches the ethics and leadership program at MIT Sloan and at other universities around the globe, including the Rotman School.

Priyadarshi asked the audience if they were happy. Then he told a short story about a man who picked him up from the airport in his new Bentley car. The man appeared very happy because of his new car and began describing its attributes. All of a sudden, he saw another Bentley coming his way, and he noted it was a newer model with more upgrades. Suddenly he didn't appear to be so happy anymore.

Tenzin asked the audience, "What is happiness?" How do we find this place of low stress? He went on to note that many of us change our cars, homes, jobs, and even spouses (sometimes a few times) in an attempt to "find happiness" and to demonstrate our life's "successes." Yet we are not satisfied. Why not?

Many of us still measure our level of success in life through our net worth. I encourage you to measure it by your self-worth. To do so requires that you find a sense of life purpose, which will go a long way toward telomerase activation and longevity. Simply put, when you have no purpose in life, your mind and body begin to die.

This chapter reinforces our need to reach out for support and advice during difficult times. If we can just "see" things differently, or get out of a stressful situation, our telomeres will not be as damaged by our stress response as they would be if we did nothing. So the goal here is to thrive when you're being pummeled by adversity–to find a higher purpose and to develop a new sense of empowerment and control over the new challenge before you.

It is possible to live a long and healthy life, but it takes work–"mindful" work. Every day, you'll need to find what drives you to want to live and to stay healthy, happy, and personally fulfilled. We will always be a work in progress. So calibrate yourself and recalibrate often. Upgrade yourself physically, mentally, and spiritually.

# 10

# Putting Your Peak Health Program Together

## Be Informed

Ever since I started researching telomeres, I have been guided by a wonderful professor who said, "Only do a test if you can do something about what you discover."

I hope this book has convinced you that you can stay healthier longer by affecting the length of your telomeres, your lifelines. You can do this by changing your lifestyle and by aggressively managing your glycation, inflammation, oxidative stress, and hormonal imbalance—using supplements, medications, and hormone replacement.

In essence, you truly can influence your genes.

The information in this book provides you with an understanding of your genetic lifeline. How you choose to abuse or preserve your lifeline will determine when you begin to fall ill—when your telomeres become critically short. And if I can state the obvious: disease often leads to death.

**Summary of Various Effects on Telomere Length**

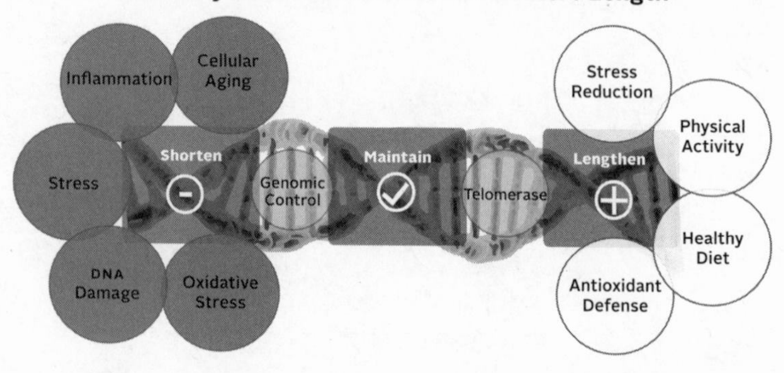


Let's use the analogy of a company bank account to bring my message home. The company has funds going into the account in the form of sales revenues. And it has funds coming out of the account to purchase goods, to provide for wages, and to pay for marketing campaigns. We all understand that a company should never spend more than it is earning, just in case there are bad days ahead (such as a global recession). And if the company does spend more than it's earning, when a crisis occurs, the company can go into debt and might go bankrupt.

Today we have the ability to determine if your health is entering a crisis management mode by testing your telomere length and identifying your percentage of critically short telomeres. If you care about the quality of your day-to-day well-being, you should become much more active in taking care of your health. You'll want to develop a basic understanding of your body's mechanisms, your aging processes, and how your lifestyle choices affect them. It probably didn't come as a surprise to you that your diet, exercise, and sleep quality are integral to your good health.

But did you know before reading this book that your habits are also influencing your telomere performance and long-term wellness? Small changes in diet can reduce your risk of the premature shortening of your telomeres and, in turn, lower your risk of multiple chronic diseases. And the most important part of living a healthy life is that you perform at your peak and enjoy your life to its fullest.

## Get Assessed

I'm often asked, "Should I be tested and learn about the length of my telomeres, or is it better if I don't know?" My answer is always "Get assessed!" Your telomere health can be managed, which means you have some control over your future.

However, in addition to testing your telomere length, I would also strongly recommend that you:

1. Assess your aging risk factors.
   a. Glycation
   b. Inflammation
   c. Oxidative stress
   d. Hormonal imbalance
2. Understand your nutrient levels.
3. Audit your eating and lifestyle habits.
4. Measure your level of physical fitness.
5. Determine your sleep quality and quantity.

When it comes to diagnostic testing, some skeptics believe that if it isn't covered by insurance or a government-paid health system, the test must either have no value or be unproven. In fact, this statement couldn't be further from the truth. Why haven't you heard about many of the tests in the list that follows? Simply put, governments and health insurers don't want to pay for them. Although they're starting to embrace the financial and economic benefits of the motto "An ounce of prevention is worth a pound of cure," our society and governments are not yet prepared to invest in this ideal. My advice is to look at some of these tests as an investment in your health.

I have divided these tests and diagnostics into three basic groups. A detailed listing of all the tests can be found at the end of the book.

1. Basic tests: these are tests you can reasonably ask your family doctor to conduct and that should be covered by your health plan. They provide a minimum baseline, which you should be receiving as part of your basic traditional checkup. These tests should be done every year.
2. Advanced tests: these tests aren't usually covered by your health plan but can be readily conducted in a traditional laboratory—for a fee, and assuming your doctor is willing to order them for you. These tests offer more insights than basic tests and will help you identify if you have any common core deficiencies. Regardless of who pays—your pocket or a third party—this type of insight into your body's health is a reasonable and worthwhile expense.
3. Professional tests: these tests require specialized laboratories. They're requisitioned by specialized health professionals who work in the field of personalized medicine. Some doctors consider these tests "unnecessary" and "superfluous," but many of them know little about the benefits and science behind the tests.

What struck me about these specialized professional tests is that they all provide well-recognized markers that can be used by various medical specialists. If I showed a cardiologist the full panel of results, she would point out just how wonderful it is to have access to the patient's cholesterol particle analysis, to a homocysteine measurement, and to coenzyme Q10 and fatty acids levels. Yet to the cardiologist, the melatonin or colon cancer risk marker might seem "unnecessary." A sleep specialist would love to correlate the melatonin levels with her brain wave studies but would have little interest in a cholesterol profile. And the cancer markers would certainly be of great interest to the gastroenterologist to determine if you would need a colonoscopy.

I am not trying to suggest you ignore your trusted health practitioner, but I'd like you to ask yourself if you are getting an opinion from someone who truly understands your health goals and the direction you'd like to follow. If you agree with the concepts of this book, you can objectively look at the tests outlined and ask for the guidance of experts who know about the subject areas. Trust is not the same as expertise. Science is objective. I've always said that numbers don't lie and facts speak loudly. The goal is to turn data into powerful personal health information and to act on that information in a very individualized way—a way that can save your life.

## Audit Your Lifestyle

Over the past twenty-five years, there is one thing I learned from medical school that has always rung true. People generally underestimate or overestimate their lifestyle habits in their favor. What do I mean by that?

### DIET

Most people underestimate the amount of food they eat. In my own practice, I see that many of my patients generally underestimate their calorie intake. When it comes to food groups, carbohydrate intake is underestimated and protein intake is overestimated. One of the most underestimated carbohydrate intakes is alcohol.

As a society, although we aim to eat sufficient servings of fruits, vegetables, and fish, few people actually follow through. Consider your own diet—and start rigorously planning your meals to include the foods that you know to be beneficial to your health.

The easiest and most effective way of analyzing a healthy diet is to log or measure your food intake. Doing this makes it possible to identify good and bad eating habits and adjust accordingly.

1. Journal: Keep track of everything you eat and drink for a week.
2. Download an app: Assess the different parts of your diet by using one of the many apps available today to log your food and beverage intake. (For example, my recommended nutrition websites include myfitnesspal.com, sparkpeople.com, and nutritiondata.self.com.)

### EXERCISE

When it comes to exercise, there are two schools of thought. There are those who already have an exercise program, one they can clearly articulate because it is part of their weekly schedule and because they have a workout routine developed in their home, office, or gym. For some in this group, there is often room for improvement; their workout level may not be optimized.

People in the other group will often say to me that they don't "work out" but that they are "active." This wishy-washy territory needs better parameters, and if you're in this group, you will benefit greatly from keeping track of your activity level for a week. Once you can clearly articulate your activity level and your commitment to your routine, it will be possible to measure the level of effort being expended to see if there is room for improvement.

1. Journal: Keep track of all your activities and their durations for a week.
2. Invest in an activity monitor: Assess the benefits of your activities by using devices such as a pedometer or an activity assessment tool (for example, Gear Fit, Fitbit, or Jawbone device).

### SLEEP

Sleep is a hugely underestimated resource in terms of body recovery and health. As a society, most of us are generally sleep deprived. Our work and family responsibilities consume so many hours of the day that, as individuals, we forfeit our sleep to meet the needs of others. We need to be realistic about the importance of sleep, and try to get more of it. The first step is understanding your current sleep patterns.

1. Journal: Keep track of your sleep and wake cycles for a week.
2. Download an app or invest in a sleep assessment device: Measure the quantity and quality of your sleep by using a monitoring device embedded in many wearable devices.

### ALCOHOL, TOBACCO, AND DRUGS

Many of us drink beer, wine, and other alcoholic drinks when we socialize with friends or with colleagues. It may seem harmless to have a drink every now and again, but the effects of regular and perhaps heavy consumption can add up and take a toll on your health. Smoking cigarettes and recreational drug use are two other common social vices, but–unlike red wine consumption, which can, under some circumstances and in very limited amounts, have a positive effect–these two habits have serious negative effects and should be cut from your routine. Be truthful about your alcohol consumption and drug use.

1. Journal: Keep track of your alcohol, tobacco, and drug use for a week.
2. Complete the alcohol, tobacco, and drug questionnaire (see Appendix).

### MENTAL HEALTH

Most of us will admit that we've endured some stressful moments in our lives. But sustained stress and an inability to cope with it can become a health issue. You might be wondering, "Am I doing OK or am I just coping?" Or perhaps you're wondering, "Am I experiencing some form of depression?"

I've always told my patients not to feel weak if they think they may be coping poorly with stress or if they are feeling depressed. An emotional injury can be equally, if not more, damaging than a physical one. Telomere studies involving stress prove this point. All too often we seek help for a sprain, migraine, or gas and bloating before we reach out for emotional help. Get some help if you *think* you need it, and certainly if you *feel* it. Complete the Goldberg Depression Questionnaire to give you some idea of how you are doing emotionally (see Appendix).

### Find Support

One of the hardest things to do is to change how we behave. According to Dr. B.J. Fogg of the Persuasive Technology Lab at Stanford University, behavioral change will only happen when three elements converge at the same moment: motivation, ability, and a trigger.[1] When behavioral change does not occur, at least one of those three elements is missing.

In medicine, all diagnostic results can act as potential triggers. How health care professionals communicate this data to you, into actionable information, is critical to triggering your own behavioral change. If I gave you a test result number but provided no context or understanding of its implications to you, you would feel no urgency to change; there would be no trigger. And behavioral change is only possible if a person's environment is conducive to that change.

Finding the time, having the money, and making a commitment to improve are critical components of your ability to change. Let me give you a few examples of some environmental situations that make it harder to change. Starting

**Changing Behavior**

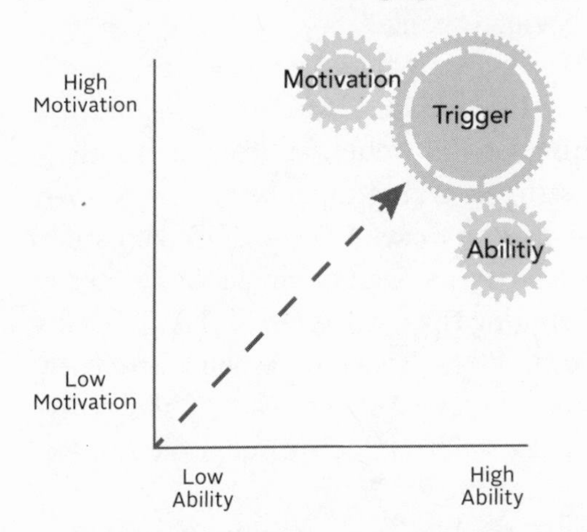


All three requirements must be met to change a behavior:

1. **motivation**
2. **ability**
3. **trigger**

The higher the motivation or ability, the more likely you'll change your behavior.

with diet, we know we should cut out all of the "bad" sugars, improve the quality of proteins, and promote a diet full of "good" fats. But if you are spending all your time on road trips and at hockey rinks or baseball diamonds, you may lack the environment to prepare healthy meals—and behavioral change is impossible.

Exercise requires a time commitment of at least several hours a week. Although some people say they need to belong to a gym to work out, I beg to differ. Lots can be done with a skipping rope and a pair of running shoes. If you work sixty hours a week or have a fistful of very young children or are caring for your aging parents, you may feel you don't have the time to commit to exercise, and this mentality pulls behavioral change out of reach. Take some time for your own health. It's essential.

Achieving an excellent quality of sleep is hard to come by if you do shift work. In fact, this type of sleep cycle practically guarantees bad sleep. Finding a different job may be difficult, but an effort needs to be made to find employment that doesn't harm your health and in an environment that provides for healthy behavioral change.

I recognize that changing habits is hard; routines are ingrained into our day-to-day lives. If we didn't have routines and habits, everything we did would feel like such a challenge because we'd struggle trying to make hundreds, even thousands, of decisions every day. A Duke University study suggests that more than 40 per cent of our actions each day are, in fact, habits.[2] Finding the motivation to change means we have to make a conscious decision to respond differently to a trigger than we have before. Work done at MIT shows that there is a simple neurological habit loop in our brains. It comprises a cue, a routine, and a reward.[3]

Here's a simple example. When you are thirsty (which is a physical discomfort cue), your routine is to get something to drink. Once you ingest some water, you are rewarded by your lack of thirst. If, however, the drink is coffee, the reward is complex. On the surface, you've satisfied your thirst, but the caffeine acts as a stimulant. Once the stimulant is metabolized, you may experience a letdown—and now you have two cues, thirst and drug-withdrawal effects, which are harder to resolve. Having too much coffee can result in a habit loop, which can result in a chemical addiction.

This simple and yet potentially complex loop highlights the importance of fully understanding your cues, routines, and rewards. One additional part of this loop is the effect of cravings. Unfortunately, cravings are often chemical in nature. Caffeine is one, and so is sugar. Alcohol, components of tobacco or drugs, and endorphins from (over-)exercising are others.

The act of responding habitually to cravings can't be eradicated because the cues in our lives will always be there. Instead, we must replace *how* we respond to them. Most often, this is our routine response. Once you identify your undesired habit loop, you can begin to find ways to replace your old routines with new ones.

As you experiment to find new routines, also try to change the reward, because rewards have a powerful effect on the cravings response. For example, many of my patients reward themselves with alcohol after a stressful day. As you know, alcohol and telomere health don't mix. Instead, how about having a nice piece of dark chocolate, which has antioxidants! By testing out different rewards, you can determine which craving you are also trying to satisfy or remove.

How do we change or break a habit? With willpower. But this often begs the next question: Am I strong enough? Does it seem like we're asking ourselves to do something strenuous? Well, in fact, we are—because willpower is like a brain muscle. If you haven't used it in a while, it needs retraining. Your willpower is strongest when you believe that you can, in fact, change. To bring about change, you must believe in yourself. Most often, that belief only emerges with the support of a group. Your odds of success improve dramatically when you commit to changing as part of a group, and change grows out of a communal experience—even if that "group" is only as large as two people.

That is why health professionals are so important to your world. They can function as health coaches, supporting you emotionally. But if your health support is distant and clinical, you might consider finding a health coach who's more engaged with your journey to better health. Lots of health care professionals can provide you with medical and factual information. But they're also there to encourage you when you think you can't do what you set out to do.

Until your new, healthier activity becomes a routine or habit, it will be difficult to sustain. Your doctor or health team is there to support you with your challenge of change.

## Get Reassessed!

Your ability to sustain a healthy lifestyle and continually set healthier goals is the most challenging part of staying healthy. Our bodies and environments are constantly in flux. We need to find ways to stay the course in the long term. There will always be setbacks, but so long as the trajectory is in the healthy direction, you are doing well. Remember: It's not where you are that's important; it's where you're headed.

Therefore, we need to find ways to reinforce our positive behaviors and prevent our recurring negative behaviors. How do we do this? By starting at the beginning again and retesting our telomeres and biomarkers to see positive changes in our metrics.

Your initial set of metrics provides direction on what areas need your prompt attention. Once you've made modifications to your lifestyle and/or taken supplements, hormones, or medications, it's time to see how much impact these changes have made on your telomeres. As we say, the numbers don't lie. Armed with a second set of measurements, you can recalibrate your action plan.

How often should you get tested? The answer is not a simple one. So let me provide you with some general advice.

Telomere testing should be done at least every six to twelve months, which will allow you to see a graphical trend. We don't want to see a downward sloping curve of your telomere length. We'd like it to stay about the same length for as long as possible. Ideally, to plot a trend, you should consider assessing your telomere length every four months, especially if you experience sudden sustained physical or emotional stress, because this can cause sudden drops in your telomere length.

Biomarkers related to disease risk, such as those related to glycation and inflammation (diabetes and cardiovascular disease), can readily improve even within one to two months. And depending on the medical concern, you might wish to reassess these types of biomarkers more frequently than every six months.

Hormones, on the other hand, respond more slowly to lifestyle changes. If you have started hormone replacement, it is important to monitor changes more frequently—say, every two to three months, to detect any hormone overdosing.

Oxidation reactions are often affected by the nutrients we ingest and by our day-to-day lifestyle. Both can vary greatly week to week, so it's best to review these more closely—say, monthly, especially if you have been advised that you have a significant nutrient deficiency. With appropriate dietary changes and supplementation, re-testing can show some steady improvement in just two to three months.

### Stay Tuned

You'll want to keep up-to-date both in areas of health you are interested in and in areas you know little about. There is so much information out there that it can be overwhelming. This doesn't mean you shouldn't try. Empowering yourself with information helps you in making a list of pros and cons on whether a behavioral change is worth it to your health and well-being. This list will help you with the "do" and "do not" decision to change. If the information is confusing, you should discuss it with a health professional who knows the area you are researching. But a word of caution: each health professional views biomarkers differently and all information comes with a bias. I have one too. I lean on the side of scientifically validated information from a more Western source, but there are many perspectives. Nobody is right all the time. Stay current, ask questions, and get answers from trusted and informed resources.

FURTHER INFORMATION

Check out a new app on iTunes called Health IQ. (Hi.Q). It offers up quizzes to test you on some of the most comprehensive measurements of your health knowledge (over 300 topics and 10,000 questions). Just a few each day will go a long way to increasing your knowledge base.

---

**Improve Your Social Network**

Health professionals can encourage you to believe in your ability to change, but it's also important to discover the sustainable support and strength of your family members, colleagues at work, and close friends within your social network.

When people hear the term *social network*, they tend to think of Facebook or Twitter. But we're all part of many networks in our day-to-day lives. Scientists have long been fascinated by how patterns of human interaction affect our individual lives. For example, why do the rich tend to get richer? One reason is that they join networks of better-off friends and gain access to opportunities to earn more. And how do you control a highly contagious disease like SARS? Epidemiologists track the social connections of infected people as they try to isolate the disease.

Nicholas Christakis and James Fowler, American scientists and authors of the bestseller *Connected*, demonstrated through their research that how we feel, what we know, whom we marry, and whether we fall ill all depend in many surprising ways on the ties that bind us. Our decisions regarding whom we connect with, how many people we connect with—and how many people *they* connect with—can subtly or dramatically influence our desires and actions. And that includes our attitude to food and physical activity. Christakis and Fowler show that a person's risk of becoming obese triples if a mutual friend becomes obese. In essence, obesity can be contagious.[4] Christakis and Fowler wrote: "We discovered that if your friend's friend's friend stopped smoking, you stopped smoking. And we discovered that if your friend's friend's friend became happy, you became happy. Eventually, we realized that there were fundamental rules."

Most of us have seen such forces at work in our lives. If you travel often on business in teams, you may have noticed how group norms form over time. If a few colleagues have three or four drinks every evening or regularly eat dessert after dinner, others feel liberated to indulge in ways they might not have if they were alone. Gradually, this becomes the norm for the whole team.

As *Connected* reveals, if most of your friends are 20 lb (10 kg) overweight, in time you will come to think of this as acceptable and may begin to gain that amount of weight as well. Conversely, if you're out of shape and are surrounded by fit people, you realize your condition is not the norm and you become motivated to be active. There is an instinctive human desire to copy other people's behavior.

I have patients who are at risk of diabetes because of their weight and lifestyle. Constantly on the road, they eat rich meals, work long hours, and don't exercise. I've come to meet some of their friends, and they, too, have become patients. Together in their social networks, they reinforce each other's bad habits. I hear the same excuse all the time: "It's just my way of life these days."

What can we do to change unhealthy behavior? Although it may be ideal to suggest they change jobs and ditch their friends to stay healthy, that's a non-starter. We need to look for ways to take control of the situation. Acknowledging that we're sliding into destructive behavior based on the influence of others is a critical first step. The second is to actively work at changing our group's norms. Take the leadership skills you muster at the office and apply them to change behavior.

Many of my patients now ensure breakfast meetings offer fruit and yogurt rather than doughnuts and muffins. They carve out time on business trips for a run or a walk, and they invite their colleagues to join. One executive who had experienced the benefits of certain supplements to keep up his energy and sustain a healthy immune system on stressful business trips decided to offer supplement packs to his executive team during a week of round-the-world town hall meetings.

I encourage you to expand your own network to include people whose habits you want to copy. Running groups, exercise buddies, peer weight-loss programs—all are good ways to tap into the positive power of social networks. And it's little wonder that the number of group activities, such as cycling teams and home yoga groups, is rising. The general population is starting to embrace the thinking behind the benefits of group health, and this can be seen in the increase of these activities.

### Find Your Purpose to Live

I've provided you with the scientific data and a guide to the available assessments to help you stay healthy. I've discussed habit loops and the importance of belief—that you *can* change. Now I give you the most difficult question to ponder.

Why do you want to live a longer and healthier life?

Over my twenty-five years of practice, and in particular the past ten years, working with executives and entrepreneurs, I have devoted my life to helping my patients understand their metrics—knowing their numbers and giving data meaning. My goal is simple: to help my patients achieve optimal health so that they can perform at their peak, at work and at play. I've helped them identify their risks of illnesses and encouraged them to change. Certainly, these patients are capable of changing their behavior and of being disciplined—they all have immense willpower, as can be seen by their successful careers.

Why do some of my patients choose *not* to change? This question has frustrated me for so many years.

Until now. I've discovered the answer. Put simply: they are not happy in their lives—either at work or at home, or both.

If you are having trouble rummaging up the willpower to change, you have to try to access your *desire* to live to see the future. You have to *want* to play with your grandchildren and enjoy your life with those close to you—your spouse, your children, your best friends. Your career has to feel like more than just a job or paycheck. You have to *desire* to make the world a better place, to contribute to your community and your social circle, and to help others—because you have something valuable to give back to the world.

The order of natural law—the law of survival of the fittest—rules in the human world too. When you have completed your evolutionary function, to procreate and pass on your genetic material to the next generation, you've achieved your biological destiny. Similarly, when you decide that you are no longer useful, the body and mind begin to wind down.

In contrast, when you are on a mission to build, contribute, and thrive with your family and society, your body rises to the occasion and enables you to do these things. There is a bounce in your steps, and you stay curious, seeking solutions to problems big and small. Your body recognizes that you not only have a biological function but a life purpose to support life beyond the self as well.

It's so important to fight and resist physical passivity and to take steps to alleviate chronic stress. My best advice is to stay active and find ways to remove yourself from chronically unhealthy environments. In other words, your decision to live is weighing the balance of external factors versus internal factors.

External factors (generally transient) include everyday things such as:

- Being in your own home
- Having a family—spouse, kids, and dog, for example
- Achieving your career aspirations, which gives you power and money
- Being financially secure and having the purchasing power to buy a house, a cottage, and cars, and to spend money on club memberships, holidays, and "things"

Once you achieve these "things," an internal voice might ask, "Is that all?" The answer is yes. In essence, you have fulfilled the basic needs of finding food and shelter for you and your offspring. It's now time to look inside for a new purpose—to live, and to live longer.

Internal factors (those that are sustainable to staying alive) include things such as:

- Having the social status, and hence the power, to change the imbalance of our society—those who have so much versus those who have so little
- Being an expert or elder in the community—empowering those in your family, co-workers, or others in your social networks to lead more useful and compassionate lives
- Leading by example to make the world a better place—being part of the global solution

Some will call these moral, social, or spiritual goals. We have a lot to learn about our mental capacity, but one thing is certain: our spiritual health is essential to our physical and mental health. I've proven this too—that happy people have longer telomeres.

Your parents provided you with a set of genetics. Your life began full of opportunities—regardless of the circumstances and environment you were born into. The research studies show us that how we carry on with our life will truly determine our health destiny. You are in control. How you have lived your life so far is in the past. Being in the present means you can do something different now. How you apply what you've learned in this book will determine what the future holds in store for you and how you live out your entire genetic potential. Getting there is all up to you.

# Afterword

## Healthy long-lifers stay true to themselves and their promises

In the opening pages of this book, I promised to share my own telomere results with you.

Along with my report, I want to explain why I think my telomeres are longer than those of other people my age, which suggests I may enjoy better health now and as I get older.

Frankly, it surprised me to learn I was above the average range for telomere length for my age group (you can see I'm above the line in the 83rd percentile). Given the stressors I experienced in childhood and through my adult life, I thought stress would get me—at the telomere level. What I think saved me was two things.

First, I've been committed to staying physically strong given a significant family history of cancer on my mother's side. This goal helped me stay true to the pillars of good health—exercise, diet, and quality sleep.

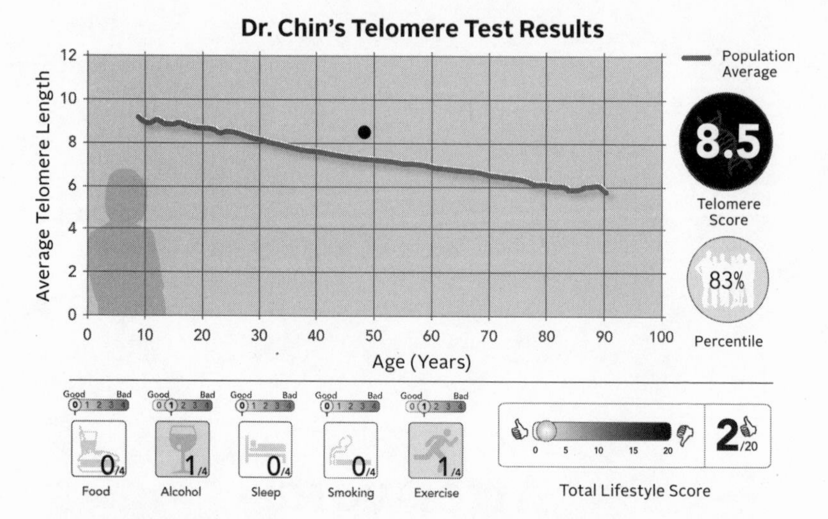


- During university and early in my marriage, I didn't exercise as much as I used to when I was younger, but I've always loved sports. Though not a great athlete, I was on the B teams for netball (a form of girls' basketball) and tennis. I'm a skier, which helps me enjoy the winter months. Today I regularly visit the gym, spinning and doing weights.
- I get enough sleep (at least seven hours during my non-training years as a physician). It was clear to me I could not function well with fewer than seven hours.
- I eat well and often at home. My mother was an amazing cook and shared her skill with me. We always had balanced meals. I had a few Big Macs from McDonald's when I was a child, and still indulge in the odd donut, when tempted, but it is never the norm for me. Today I enjoy eating all types of food. I cook at home a lot and eat out in healthy restaurants. My meals are balanced: lots of grains, vegetables, and fruits with meat, fish, and fowl. I eat real food (always try to avoid the processed stuff). I never skip a meal. Nor do I snack that often.
- As for the bad habits that can shorten telomeres, I drink alcohol socially occasionally, a few nights a month. But I never

overindulge (no more than two glasses). Like most Chinese people, I don't have the enzyme to break down alcohol. That genetic factor makes me a "cheap date" and often the designated driver. And I don't smoke—except for the one draw of a cigarette I admitted to at the beginning of this book that turned me off forever, thankfully.

Second, I think that the most significant contributor to a healthy telomere result is the attitude I adopt even when dealing with personal setbacks. I describe the importance of mental resilience in Chapter 9. Resilience comes from having a desire to live, a life purpose, and never giving up believing in yourself.

What drives me? My life purpose is to help people stay healthy so that they can perform at their peak—physically and emotionally. That's why I wrote this book—to summarize my lifelong learning to help you and me and others to enjoy good health; and to help people avoid bad lifestyle behaviors and environmental traumas that can lead to premature death. It's my hope that this book will help readers like you achieve peak performance right to the tips of your telomeres—for a very long time.

### Some advice about success and finding purpose in life from Don Morrison

The following is written by a global leader and dear friend, Don Morrison. His note complements my encouragement of you to manage not only your physical and mental well-being but also your spiritual being. He has taught me the importance of practicing altruism and compassion in our everyday life.

> One bright winter day, on my first day of work in January of 1999, I stood in my new office on the top floor of a downtown skyscraper in Toronto. I had just arrived at my new job

as president of consumer and small business for AT&T Canada. The past twenty-plus years had taken me and my family from my early career days in Toronto, to Washington, D.C., and then to Europe, the Middle East, and Africa, and ultimately back home. As I stood there looking out the window, having come to the realization that I had achieved, at least in my profession, the very success that I had worked so hard for, I asked myself, "Is that all there is?" Perhaps you have had a similar moment in your life.

I see more clearly the answer to this question now, fifteen years later, than I did on that cold wintery day. Yet, although the question is clearer, with respect to the answer, I still see that I am on a journey of discovery.

What I see as I reflect today is a coexistence of two very complementary things. On the one hand, the joy of achievement I felt was very much attached to my "external journey," that is, my business career, my external identity. On the other hand, any angst that I felt that day was due to my deep inner knowledge that this success alone was not sufficient, that there must be a corresponding "inner journey" that is neither measured nor understood in the same manner as the exterior material journey. This inner journey still lay ahead of me.

Regarding this external journey, many self-help books will tell you that you should feel guilty for your success. They will justify their position by characterizing the fleeting satisfaction and seeming superficiality of success as frivolous when compared to other pursuits. For some reason, they juxtapose the idea of success—the way we define it—with other things of deeper intrinsic value, like purpose, meaning, and selflessness.

Personally, I do not see this external characterization of success as contradicting deeper questions, but more as complementing them. I have great difficulty relating to an education system and a culture that encourage you to be a success, and then the moment that you begin to break out and really make something out of your life, make you feel guilty for it. It seems as though there is a force in our culture that seeks to pull everything toward a magnetic pole of mediocrity.

The truth is that it is far more fulfilling, for both the individual and for society in general, to encourage all of us to be the very best that we can be in our chosen profession. This is really the first journey in life, focusing on those "external factors," as Dr. Elaine Chin refers to them in this book.

But life is not just about accumulating things like material wealth. Unfortunately, if you look at what our modern culture values today, you can draw the conclusion that everything depends on what you wear, who you are, whom you know, what you own, etc.

However, there is a second journey in life, one that starts a little later than the first journey, and runs parallel with it. This second journey is the inner, deeper journey that deals with the cultivation of the "true self." Just as the first journey is about acquiring things like material wealth, and an identity or a reputation, the second journey is more about letting go. The first journey is about building up. The second journey is about taking down. It is about seeing the illusion of that external identity that you have built up in your mind. It is about finding time for silence. It is about embarking on a journey toward a deeper understanding of why we are here on this planet for this brief period of time. Despite what others might tell you, this is not about building a legacy; the issue of legacy is perhaps the last chapter of the external journey.

This inner journey is traversed on the path of paradox. By this I mean that the great irony of the search to find oneself is to come to the realization that this search really has nothing to do with the self at all. In fact, one discovers that the self as we know it is really an illusion that we have created in our own minds to help us explain ourselves to ourselves so that we can survive day-to-day in the marketplace.

So, why embark on this inner journey in the first place? The simple answer is that this question lies at the very heart of why you are alive at this moment. No one can answer the question for you as to who you truly are, but anyone who has written authoritatively on this subject will tell you, in another twist of irony, that the very thing you need to discover is something that you already know. You have to just scrape away all of the accumulated ego and identity, and fear and ambition, and agendas and plans, and your need to control in order to "see" *you* more clearly. The closest we can come to having guides, other than another person to actually guide us, is to reflect on art, poetry, and music. Somewhere, between the lines, or on a canvas, or in a melody, lies a hidden wholeness that may speak to us about our own existence. Or it may lie in a deep unknowing that is cultivated through the daily practice of meditation.

The consequence is a new insight that renders a more wise and fulfilled consciousness in which the individual seeks to identify themselves, beyond themselves, in some manner that betters mankind.

This path isn't for everyone. That is why it is often referred to as the "road less traveled." But if a person embarks with pure intention on this inner journey, the benefits are immeasurable, because they awaken to a new way of seeing, and a renewed energy, that makes their life worth living as it affects everyone and everything around them.

DONALD H. MORRISON

Since retiring as the Chief Operating Officer of Research in Motion, Don has begun work on projects of importance to him. He is a Director of the MasterCard Foundation, a member of the Dalai Lama Center for Ethics and Transformative Values at MIT, the advisory board of the Mind and Life Institute in Canada, the Centre for International Governance Innovation (CIGI) and has recently commissioned a project to develop new methods for the cultivation and practice of altruism and compassion with the Center for Engaged Compassion at Claremont College, California.

# Appendix: Diagnostic Laboratory Tests Checklist

**Basic Tests**

1. Glucose
2. Cholesterol (total and triglycerides)
3. Organ system function for:
   a. Liver
      - AST (Aspartate transaminase)
      - ALT (Alanine transferase)
      - ALP (Alkaline phosphatase)
      - GGT (Gamma-glutamyl transpeptidase)

   b. Kidney
      - Creatinine

   c. Immune system
      - Hemoglobin
      - White blood cells
4. Vital signs
   - Percentage body water
   - Weight
   - Blood pressure

- Heart rate
- Lung–aerobic capacity
- Percentage body weight
- Percentage visceral fat
- Lean body mass

### Advanced Tests

1. Hemoglobin A1C

2. Cholesterol profiling
    - LDL and subparticles
    - HDL and subparticles
    - Lipoproteins

3. Hormone panel
    - Estrogens (estradiol, estriol, estrone)
    - Progesterone
    - Testosterone
    - Dehydroepiandrosterone (DHEA)
    - Thyroid function (TSH–thyroid-stimulating hormone, T4, and T3)
    - Insulin

### Professional Tests

1. Telomere test
    - Average length
    - Percentage short telomeres

2. Genetics
    - Pharmacogenetics
    - Nutrigenomics
    - Disease risk–population genetics
    - Mutations

3. DNA damage levels
    - 8-hydroxy-2'-deoxyguanosine

4. Inflammation
    - C-reactive protein
    - Homocysteine
    - Arachidonic acid
    - Prostaglandins
    - Interleukins
    - Omega fatty acids

5. Oxidation panel
    - Vitamin C
    - Coenzyme Q10
    - Superoxide dismutase
    - Catalase
    - Lipid peroxidase
    - Glutathione peroxidase

6. Hormone panel
    - Cortisol
    - Growth hormone
    - Serotonin
    - Melatonin

7. Nutrient panel
    - Vitamins
    - Minerals
    - Amino acids
    - Fatty acids
    - Organic acids
    - Toxin panel
    - Mercury
    - Lead
    - Arsenic
    - Cadmium

## Drug Abuse Screening Test (DAST-20)



The following questions concern information about your potential involvement with drugs not including alcoholic beverages during the past twelve months. Carefully read each statement and decide if your answer is "yes" or "no." Then, circle the appropriate response beside the question.

In the statements "drug abuse" refers to (1) the use of prescribed or over the counter drugs in excess of the directions and (2) any non-medical use of drugs. The various classes of drugs may include: cannabis (e.g. marijuana, hash), solvents, tranquilizers (e.g. Valium), barbiturates, cocaine, stimulants (e.g. speed), hallucinogens (e.g. LSD) or narcotics (e.g. heroin). Remember that the questions do not include alcoholic beverages.

Please answer every question. If you have difficulty with a statement, then choose the response that is mostly right.

These questions refer to the past twelve months. Answer with yes or no.

1. Have you used drugs other than those required for medical reasons?
2. Have you abused prescription drugs?
3. Do you abuse more than one drug at a time?
4. Can you get through the week without using drugs?
5. Are you always able to stop using drugs when you want to?
6. Have you had "blackouts" or "flashbacks" as a result of drug use?
7. Do you ever feel bad or guilty about your drug use?
8. Does your spouse (or parents) ever complain about your involvement with drugs?

9. Has drug abuse created problems between you and your spouse or your parents?
10. Have you lost friends because of your use of drugs?
11. Have you neglected your family because of your use of drugs?
12. Have you been in trouble at work because of drug abuse?
13. Have you lost a job because of drug abuse?
14. Have you gotten into fights when under the influence of drugs?
15. Have you engaged in illegal activities in order to obtain drugs?
16. Have you been arrested for possession of illegal drugs?
17. Have you ever experienced withdrawal symptoms (felt sick) when you stopped taking drugs?
18. Have you had medical problems as a result of your drug use (e.g. memory loss, hepatitis, convulsions, bleeding, etc.)?
19. Have you gone to anyone for help for a drug problem?
20. Have you been involved in a treatment program specifically related to drug use?

SCORING:

A factor analysis of the twenty items has indicated that the DAST is essentially a uni-dimensional scale. Accordingly, it is planned to yield only one total or summary score ranging from 0 to 20, which is computed by summing all items that are endorsed in the direction of increased drug problems. Only two items are keyed for a "no" response: "Can you get through the week without using drugs?" and "Are you always able to stop using drugs when you want to?" A DAST score of 6 or above is suggested for case finding purposes, since most of the clients in the normative sample score 6 or greater. It is also suggested that a score of 16 or greater be considered to indicate a very severe abuse or a dependency condition.

**Goldberg Depression Questionnaire, a Screening Test for Depression**

Answer the following questions with:

- Not at all
- Just a little
- Somewhat
- Moderately
- Quite a lot
- Very much

1. I do things slowly.
2. My future seems hopeless.
3. It is hard for me to concentrate on reading.
4. The pleasure and joy have gone out of my life.
5. I have difficulty making decisions.
6. I have lost interest in aspects of my life that used to be important to me.
7. I feel sad, blue, and unhappy.
8. I am agitated and keep moving around.
9. I feel fatigued.
10. It takes great effort for me to do simple things.
11. I feel that I am a guilty person who deserves to be punished.
12. I feel like a failure.
13. I feel lifeless, more dead than alive.
14. I'm getting too much, too little, or not enough restful sleep.
15. I spend time thinking about *how* I might kill myself.
16. I feel trapped or caught.
17. I feel depressed even when good things happen to me.
18. Without trying to diet, I have lost or gained weight.

SCORING:

- Not at all: 0 points
- Just a little: 1 point
- Somewhat: 2 points
- Moderately: 3 points
- Quite a lot: 4 points
- Very much: 5 points

SCREENING TEST SCORING RANGES:

- 0–9, No Depression Likely
- 10–17, Possibly Mildly Depressed
- 18–21, Borderline Depression
- 22–35, Mild–Moderate Depression
- 36–53, Moderate–Severe Depression
- 54 and up, Severely Depressed

The higher the number, the more severe the depression. If you take the quiz again weekly or monthly, changes of 5 or more points between tests may be significant.

# Notes

## 2 | Welcome to the New Science of Aging


1 Marsa, Linda. "Scientist of the year notable: Elizabeth Blackburn." *Discover Magazine*. December 6, 2007. Available at discovermagazine.com/2007/dec/blackburn#.UwOJ3-tKWsZ. Accessed February 18, 2014.

2 Egan, Tom. "New views on an age-old question." *The Guardian*. May 13, 2002. Available at guardian.co.uk/theguardian/2002/may/13/guardianletters. Accessed February 18, 2014.

3 http://med.stanford.edu/news/all-news/2015/01/telomere-extension-turns-back-aging-clock-in-cultured-cells.html

4 Page, Larry. "Google announces Calico, a new company focused on health and well-being." [Google website]. September 18, 2013. Available at googlepress.blogspot.ca/2013/09/calico-announcement.html. Accessed February 18, 2014.

5 Buettner, Dan. *The Blue Zone: Lessons for Living Longer from the People Who've Lived the Longest*. Washington, DC: National Geographic Society; 2008.

6 Slagboom, Eline, Droog S, Boomsma DI. "Genetic determination of telomere size in humans: a twin study of three age groups." *American Journal of Human Genetics*. 1994 Nov; 55(5): 876–82.

7 Huda, Nazmul, Tanaka H, Herbert BS, Reed T, Gilley D. "Shared environmental factors associated with telomere length maintenance in elderly male twins." *Aging Cell.* 2007 Oct; 6(5): 709–13.

8 Panossian, LA, Porter VR, Valenzuela HF, Zhu X, Reback E, Masterman D, Cummings JL, Effros, RB. "Telomere shortening in T cells correlates with Alzheimer's disease status." *Neurobiology of Aging.* 2003 Jan-Feb; 24(1): 77–84.

9 Honig, Lawrence S., Kang MS, Schupf N, Lee JH, Mayeux R. "Association of shorter leukocyte telomere repeat length with dementia and mortality." *Arch Neurol.* 2012 Oct; 69(10): 1,332–39.

10 Minamino, Tohru, Miyauchi H, Yoshida T, Ishida Y, Yoshida H, Komuro I. "Endothelial cell senescence in human atherosclerosis: role of telomere in endothelial dysfunction." *Circulation.* 2002 Apr 2; 105(13): 1,541–44.

11 Fitzpatrick, Annette L., Kronmal RA, Gardner JP, Psaty BM, Jenny NS, Tracy RP, Walston J, Kimura M, Aviv A. "Leukocyte telomere length and cardiovascular disease in the cardiovascular health study." *Am J Epidemiol.* 2007 Jan 1; 165(1): 14–21.

12 Brouilette, Scott, Singh RK, Thompson JR, Goodall AH, Samani NJ. "White cell telomere length and risk of premature myocardial infarction." *Arterioscler Thromb Vasc Biol.* 2003 May 1; 23(5): 842–6. Epub 2003 Mar 20.

13 von Zglinicki, Thomas, Serra V, Lorenz M, Saretzki G, Lenzen-Grossimlighaus R, Gessner R, Risch A, Steinhagen-Thiessen E. "Short telomeres in patients with vascular dementia: an indicator of low antioxidative capacity and a possible risk factor?" *Lab Invest.* 2000 Nov; 80(11): 1,739–47.

14 Blasco, Maria A. "Telomeres and cancer: a tale with many endings." *Curr Opin Genet Dev.* 2003 Feb; 13(1): 70–6.

15 Cawthon, Richard M., Smith KR, O'Brien E, Sivatchenko A, Kerber RA. "Association between telomere length in blood and mortality in people aged 60 years or older." *Lancet.* 2003 Feb 1; 361(9355): 393–5.

16 Kitada, T, Seki S, Kawakita N, Kuroki T, Monna T. "Telomere shortening in chronic liver diseases." *Biochem Biophys Res Commun.* 1995 Jun 6; 211(1): 33–9.

17 Harbo, Maria, Bendix L, Bay-Jensen AC, Graakjaer J, Søe K, Andersen TL, Kjaersgaard-Andersen P, Koelvraa S, Delaisse JM. "The distribution pattern of critically short telomeres in human osteoarthritic knees." *Arthritis Res Ther.* 2012 Jan 18; 14(1): R12. doi: 10.1186/ar3687.

18 Zhai, G, Aviv A, Hunter DJ, Hart DJ, Gardner JP, Kimura M, Lu X, Valdes AM, Spector TD. "Reduction of leucocyte telomere length in radiographic hand osteoarthritis: a population-based study." *Ann Rheum Dis.* 2006 Nov; 65(11): 1,444–48.

19 Saeed, Hamid, Abdallah BM, Ditzel N, Catala-Lehnen P, Qiu W, Amling M, Kassem M. "Telomerase-deficient mice exhibit bone loss owing to defects in osteoblasts and increased osteoclastogenesis by inflammatory microenvironment." *J Bone Miner Res.* 2011 Jul; 26(7): 1,494–505. doi: 10.1002/jbmr.349.

20 Valdes, AM, Richards JB, Gardner JP, Swaminathan R, Kimura M, Xiaobin L, Aviv A, Spector TD. "Telomere length in leukocytes correlates with bone mineral density and is shorter in women with osteoporosis." *Osteoporos Int.* 2007 Sep; 18(9): 1,203–10.

21 Kosmadaki, MG, Gilchrest BA. "The role of telomeres in skin aging/photoaging." *Micron.* 2004; 35(3): 155–9.

22 Nakagawa, Shinichi, Gemmell NJ, Burke T. "Measuring vertebrate telomeres: applications and limitations." *Mol Ecol.* 2004 Sep; 13(9): 2,523–33.

23 Haussmann, Mark. "Sea birds' DNA may hold keys to aging and cancer, researcher says." Available at bucknell.edu/x45446.xml. Accessed January 27, 2015.

24 Harley, Calvin, and Epel, Elissa. "Telomeres and Aging: how stress causes aging." Filmed 2011. TedMed Talk, 12:41. Posted January 2012. youtube.com/watch?v=mnMuuRcczf8

## 3 | What Are Telomeres?

1 Hemann, Michael T, Strong MA, Hao LY, Greider CW. "The shortest telomere, not average telomere length, is critical for cell viability and chromosome stability." *Cell.* 2001 Oct 5; 107(1): 67–77.

2 Bodnar, Andrea G, Ouellette M, Frolkis M, Holt SE, Chiu CP, Morin GB, Harley CB, Shay JW, Lichtsteiner S, Wright WE. "Extension of lifespan by introduction of telomerase into normal human cells." *Science.* 1998 Jan 16; 279(5349): 349–52.

3 Fossel, Michael. "Telomerase and the aging cell: implications for human health." *Journal of the American Medical Association.* 1998 Jun 3; 279(21): 1,732–35.

4 Jaskelioff, Mariela, Muller FL, Paik JH, Thomas E, Jiang S, Adams AC, Sahin E, Kost-Alimova M, Protopopov A, Cadiñanos J, Horner JW, Maratos-Flier E, Depinho RA. "Telomerase reactivation reverses tissue degeneration in aged telomerase-deficient mice." *Nature*. 2011 Jan 6; 469(7328): 102–6. doi: 10.1038/nature09603.

## 4 | Know Your Biomarkers and the Aging Processes

1 Giampapa, Vincent C. *The Principles and Practice of Antiaging Medicine for the Clinical Physician*. Denmark: River Publishers; 2012.

2 Glassman, RH, Ratain MJ. "Biomarkers in early cancer drug development: limited utility." *Clin Pharmacol Ther.* 2009 Feb; 85(2): 134–5. doi: 10.1038/clpt.2008.231.

3 Ryan, G, Knuiman MW, Divitini ML, James A, Musk AW, Bartholomew HC. "Decline in lung function and mortality: the Busselton Health Study." *J Epidemiol Community Health*. 1999 Apr; 53(4): 230–4.

4 Krogsbøll, Lasse T, Jørgensen KJ, Grønhøj Larsen C, Gøtzsche PC. "General health checks in adults for reducing morbidity and mortality from disease: Cochrane systematic review and meta-analysis." *BMJ*. 2012 Nov 20; 345: e7191. doi: 10.1136/bmj.e7191.

5 Gupta S. The Last Heart Attack. [CNN Health website]. August 29, 2011. Available at sanjayguptamd.blogs.cnn.com/2011/08/29/sanjay-gupta-reports-the-last-heart-attack

6 Jeanclos, E, Krolewski A, Skurnick J, Kimura M, Aviv H, Warram JH, Aviv A. "Shortened telomere length in white blood cells of patients with IDDM." *Diabetes*. 1998 Mar; 47(3): 482–6.

7 Gardner, Jeffrey P, Li S, Srinivasan SR, Chen W, Kimura M, Lu X, Berenson GS, Aviv A. "Rise in insulin resistance is associated with escalated telomere attrition." *Circulation*. 2005 May 3; 111(17): 2,171–77.

8 Kim, Sangmi, Parks CG, DeRoo LA, Chen H, Taylor JA, Cawthon RM, Sandler DP. "Obesity and weight gain in adulthood and telomere length." *Cancer Epidemiol Biomarkers Prev.* 2009 Mar; 18(3): 816–20. doi: 10.1158/1055-9965.EPI-08-0935.

9 Zannolli, Raffaella, Mohn A, Buoni S, Pietrobelli A, Messina M, Chiarelli F, Miracco C. "Telomere length and obesity." *Acta Paediatr.* 2008 Jul; 97(7): 952–4. doi: 10.1111/j.1651-2227.2008.00783.x.

10 Chen, Wei, Gardner JP, Kimura M, Brimacombe M, Cao X, Srinivasan SR, Berenson GS, Aviv A. "Leukocyte telomere length is associated

with HDL cholesterol levels: The Bogalusa heart study." *Atherosclerosis.* 2009 Aug; 205(2): 620–5.

11 Pfützner, Andreas, and Forst, Thomas. "High-sensitivity C-reactive protein as cardiovascular risk marker in patients with diabetes mellitus." *Diabetes Technol* Ther. 2006 Feb; 8(1): 28–36.

12 Richards, JB, Valdes AM, Gardner JP, Kato BS, Siva A, Kimura M, Lu X, Brown MJ, Aviv A, Spector TD. "Homocysteine levels and leukocyte telomere length." *Atherosclerosis.* 2008 Oct; 200(2): 271–7. doi: 10.1016/j.atherosclerosis.2007.12.035.

13 Vitamin C: Fact Sheet for Health Professionals. [National Institutes of Health, Office of Dietary Supplements]. Available at ods.od.nih.gov/factsheets/VitaminC-HealthProfessional/. Accessed February 18, 2014.

## 5 | Tune Up Your Diet

1 Stoll, LL, McCormick ML, Denning GM, Weintraub NL. "Antioxidant effects of statins." *Timely Top Med Cardiovasc Dis.* 2005 Feb 4; 9: E1.

2 "Sugar: The Bitter Truth," YouTube video from a performance televised on July 30, 2009, youtube.com/watch?v=dBnniua6-oM. University of California TV, 1:29. Posted June 2009. Accessed April 10, 2014.

3 Body mass index, overweight or obese, self-reported, adult, by age group and sex. [Statistics Canada website]. Available at statcan.gc.ca/tables-tableaux/sum-som/l01/cst01/health81a-eng.htm. Accessed February 19, 2014.

4 A By-the-Numbers Look at Obese Children. [CBC website]. Available at cbc.ca/news2/interactives/obesity-children-canada/index.html. Accessed April 10, 2014.

5 U.S. Obesity Rate Climbing in 2013. [Gallup Well-Being website]. Available at gallup.com/poll/165671/obesity-rate-climbing-2013.aspx. Accessed February 19, 2014.

6 Clinton, Bill. *Giving: How Each of Us Can Change the World.* New York: Knopf; 2007: 37–38.

7 Diagnosing Diabetes and Learning About Prediabetes. [American Diabetes Association website]. Available at diabetes.org/diabetes-basics/diagnosis/. Accessed February 19, 2014.

8 Definition, Classification and Diagnosis of Diabetes, Prediabetes and Metabolic Syndrome [Canadian Diabetes Association website]. Accessed April 10, 2014.

9 Economic Costs of Diabetes in the U.S. in 2012. [American Diabetes Association website]. Available at care.diabetesjournals.org/content/36/4/1033.full. Accessed February 19, 2014.

10 The prevalence and costs of diabetes. [Canadian Diabetes Association website]. Available at diabetes.ca/diabetes-and-you/what/prevalence/. Accessed February 19, 2014.

11 Sclafani, Anthony, and Springer, Deleri. "Dietary obesity in adult rats: similarities to hypothalamic and human obesity syndromes." *Physiol Behav.* 1976 Sep; 17(3): 461–71.

12 Moss, Michael. *Salt, Sugar, Fat: How the Food Giants Hooked Us*. New York: Random House; 2013.

13 Carden, Trevor J., and Carr, Timothy P. "Food availability of glucose and fat, but not fructose, increased in the U.S. between 1970 and 2009: analysis of the USDA food availability data system." *Nutr J.* 2013; 12: 130. doi: 10.1186/1475-2891-12-130.

14 Stanhope, Kimber, Bremer AA, Medici V, Nakajima K, Ito Y, Nakano T, Chen G, Fong TH, Lee V, Menorca RI, Keim NL, Havel PJ. "Consumption of fructose and high fructose corn syrup increase postprandial triglycerides, LDL-cholesterol, and apolipoprotein-B in young men and women." *J Clin Endocrinol Metab.* 2011 Oct; 96(10): E1, 596–605. doi: 10.1210/jc.2011-1251. Epub 2011 Aug 17.

15 The Warnings from a Cancer Doctor. [Fifth Estate website]. Available at cbc.ca/fifth/blog/the-warnings-from-a-cancer-doctor. Accessed February 19, 2014.

16 Preview: Is Sugar Toxic? Available at cbsnews.com/news/preview-is-sugar-toxic/. Accessed February 19, 2014.

17 Davidson, Terry L., Martin AA, Clark K, Swithers SE. "Intake of high-intensity sweeteners alters the ability of sweet taste to signal caloric consequences: implications for the learned control of energy and body weight regulation." *Q J Exp Psychol (Hove).* 2011 Jul; 64(7): 1,430–41.

18 Haffner, Steven M. "Obesity and the metabolic syndrome: the San Antonio Heart Study." *Br J Nutr.* 2000 Mar; 83 Suppl 1: S67–70.

19 Schulze, Matthias B., Manson JE, Ludwig DS, Colditz GA, Stampfer MJ, Willett WC, Hu FB. "Sugar-sweetened beverages, weight gain, and incidence of type 2 diabetes in young and middle-aged women." *JAMA*, 2004 Aug; 292(8): 927–34.

20 Rolls, Barbara J. "Effects of intense sweeteners on hunger, food intake, and body weight: a review." *Am J Clin Nutr.* 1991 Apr; 53(4): 872–8.

21 Smith, Kylie J., Sanderson K, McNaughton SA, Gall SL, Dwyer T, Venn AJ. "Longitudinal associations between fish consumption and depression in young adults." *Am J Epidemiol.* 2014 May 15; 179(10): 1,228–35.

22 Chein, Edmund, and Hiroshi, Demura H. *Bio-Identical Hormones and Telomerase: The Nobel Prize–Winning Research Into Human Life Extension and Health.* iUniverse.com; 2011.

23 Mann, Traci, Tomiyama AJ, Westling E, Lew AM, Samuels B, Chatman J. "Medicare's search for effective obesity treatments: diets are not the answer." *Am Psychol.* 2007 Apr; 62(3): 220–33.

24 The Biosphere 2 Project–A Laboratory for Global Ecology. [BioSpherics website]. Available at biospherics.org/biosphere2/results/1-the-biosphere-2-project-a-laboratory-for-global-ecology/. Accessed April 10, 2014.

## 6 | Supplements to Counteract Your Deficiencies

1 Krebs-Smith, SM, Cleveland LE, Ballard-Barbash R, Cook DA, Kahle LL. "Characterizing food intake patterns of American adults." *Am J Clin Nutr.* 1997 Apr; 65(4 Suppl): 1,264S–1,268S.

2 Do Canadian Adults Meet Their Nutrient Requirements Through Food Intake Alone? [Health Canada website]. Available at hc-sc.gc.ca/fn-an/surveill/nutrition/commun/art-nutr-adult-eng.php. Accessed February 19, 2014.

3 Lonn, Eva, Bosch J, Yusuf S, Sheridan P, Pogue J, Arnold JM, Ross C, Arnold A, Sleight P, Probstfield J, Dagenais GR; HOPE and HOPE-TOO Trial Investigators. "Effects of long-term vitamin E supplementation on cardiovascular events and cancer: a randomized controlled trial." *JAMA.* 2005 Mar 16; 293(11): 1,338–47.

4 Sanders, Kerrie M, Stuart AL, Williamson EJ, Simpson JA, Kotowicz MA, Young D, Nicholson GC. "Annual high-dose oral vitamin D and falls and fractures in older women: a randomized controlled trial." *JAMA.* 2010 May 12; 303(18): 1,815–22. doi: 10.1001/jama.2010.594.

5 Harris, William S. "Fish oils and plasma lipid and lipoprotein metabolism in humans: a critical review." *J Lipid Res.* 1989 Jun; 30(6): 785–807.

6 Brasky, Theodore M., Darke AK, Song X, Tangen CM, Goodman PJ, Thompson IM, Meyskens FL Jr, Goodman GE, Minasian LM, Parnes HL, Klein EA, Kristal AR. "Plasma phospholipid fatty acids and prostate cancer risk in the SELECT trial." *Natl Cancer Inst.* 2013 Aug 7; 105(15): 1,132–41. doi: 10.1093/jnci/djt174.

7 Farzaneh-Far, Ramin, Lin J, Epel ES, Harris WS, Blackburn EH, Whooley MA. "Association of marine omega-3 fatty acid levels with telomeric aging in patients with coronary heart disease." *JAMA.* 2010 Jan 20; 303(3): 250–7. doi: 10.1001/jama.2009.2008.

8 Xu, Qun, Parks CG, DeRoo LA, Cawthon RM, Sandler DP, Chen H. "Multivitamin use and telomere length in women." *Am J Clin Nutr.* 2009 Jun; 89(6): 1,857–63. doi: 10.3945/ajcn.2008.26986.

9 Richards, J Brent, Valdes AM, Gardner JP, Paximadas D, Kimura M, Nessa A, Lu X, Surdulescu GL, Swaminathan R, Spector TD, Aviv A. "Higher serum vitamin D concentrations are associated with longer leukocyte telomere length in women." *Am J Clin Nutr.* 2007 Nov; 86(5): 1,420–25.

10 Zhang, Wei-Jian, Hufnagl P, Binder BR, Wojta J. "Antiinflammatory activity of astragaloside IV is mediated by inhibition of NF-kappaB activation and adhesion molecule expression." *Thromb Haemost.* 2003 Nov; 90(5): 904–14.

## 7 | The Power of Hormone-Replacement Therapy

1 Epel, Elissa S. "Psychological and metabolic stress: a recipe for accelerated cellular aging?" *Hormones* (Athens). 2009 Jan–Mar; 8(1): 7–22.

2 Barbieri, Michaelangela, Paolisso G, Kimura M, Gardner JP, Boccardi V, Papa M, Hjelmborg JV, Christensen K, Brimacombe M, Nawrot TS, Staessen JA, Pollak MN, Aviv A. "Higher circulating levels of IGF-1 are associated with longer leukocyte telomere length in healthy subjects." *Mech Ageing Dev.* 2009 Nov–Dec; 130(11-12): 771–6. doi: 10.1016/j.mad.2009.10.002.

3 Sherwin, BB. "Sex hormones and psychological functioning in postmenopausal women." *Exp Gerontol.* 1994 May-Aug; 29(3–4): 423–30.

4 Findings from the WHI Postmenopausal Hormone Therapy Trials. [Women's Health Initiative website]. Available at www.nhlbi.nih.gov/whi/. Accessed February 19, 2014.

5 Jacobs, Emily G, Kroenke C, Lin J, Epel ES, Kenna HA, Blackburn EH, Rasgon NL. "Accelerated cell aging in female APOE-ϵ4 carriers:

implications for hormone therapy use." *PLoS One.* 2013; 8(2): e54713. doi: 10.1371/journal.pone.0054713.

6 Tang, Ming-Xin, Jacobs D, Stern Y, Marder K, Schofield P, Gurland B, Andrews H, Mayeux R. "Effect of oestrogen during menopause on risk and age at onset of Alzheimer's disease." *Lancet.* 1996 Aug 17; 348(9025): 429–32.

7 Misiti, Silvia, Nanni S, Fontemaggi G, Cong YS, Wen J, Hirte HW, Piaggio G, Sacchi A, Pontecorvi A, Bacchetti S, Farsetti A. "Induction of hTERT expression and telomerase activity by estrogens in human ovary epithelium cells." *Mol Cell Biol.* 2000 Jun; 20(11): 3,764–71.

8 Williams, Christopher D, Boggess JF, LaMarque LR, Meyer WR, Murray MJ, Fritz MA, Lessey BA. "A prospective, randomized study of endometrial telomerase during the menstrual cycle." *J Clin Endocrinol Metab.* 2001 Aug; 86(8): 3,912–77.

9 Saad, Farid. "Androgen therapy in men with testosterone deficiency: can testosterone reduce the risk of cardiovascular disease?" *Diabetes Metab Res Rev.* 2012 Dec; 28 Suppl 2: 52–9. doi: 10.1002/dmrr.2354.

10 Janowsky, Jeri S, Oviatt SK, Orwoll ES. "Testosterone influences spatial cognition in older men." *Behav Neurosci.* 1994 Apr; 108(2): 325–32.

11 Barqawi A, and Crawford ED. "Testosterone replacement therapy and the risk of prostate cancer. Is there a link?" *Int J Impot Res.* 2006 Jul–Aug; 18(4): 323–8.

12 Cho, Stephen. "Prostate Cancer Risk Not Elevated with Testosterone Replacement." [Renal & Eurology News website]. Available at renalandurologynews.com/prostate-cancer-risk-not-elevated-with-testosterone-replacement/article/292346/. Accessed February 19, 2014.

13 Khaw, Kay-Tee, Dowsett M, Folkerd E, Bingham S, Wareham N, Luben R, Welch A, Day N. "Endogenous testosterone and mortality due to all causes, cardiovascular disease, and cancer in men: European prospective investigation into cancer in Norfolk (EPIC-Norfolk) Prospective Population Study." *Circulation.* 2007 Dec 4; 116(23): 2, 694–701. Epub 2007 Nov 26.

14 Morley, JE, Kaiser FE, Sih R, Hajjar R, Perry HM 3rd. "Testosterone and frailty." *Clin Geriatr Med.* 1997 Nov; 13(4): 685–95.

15 Björntorp, Per. "Hormonal control of regional fat distribution." *Hum Reprod.* 1997 Oct; 12 Suppl 1:21–5.

16 Kapoor, D, Goodwin E, Channer KS, Jones TH. "Testosterone replacement therapy improves insulin resistance, glycaemic control, visceral adiposity and hypercholesterolaemia in hypogonadal men with type 2 diabetes." *Eur J Endocrinol.* 2006 Jun; 154(6): 899–906.

17 Phillips, Gerald B, Pinkernell BH, Jing TY. "The association of hypotestosteronemia with coronary artery disease in men." *Arterioscler Thromb.* 1994 May; 14(5): 701–6.

18 Meeker, AK, Sommerfeld HJ, Coffey DS. "Telomerase is activated in the prostate and seminal vesicles of the castrated rat." *Endocrinology.* 1996 Dec; 137(12): 5,743–46.

19 Basson R. "Testosterone therapy for reduced libido in women." *Ther Adv Endocrinol Metab.* 2010 Aug; 1(4): 155–64. doi: 10.1177/2042018810379588.

20 Shealy, C. *DHEA: The Youth and Health Hormone.* New York City: McGraw-Hill Education; 1999.

21 Gómez-García, L, Sánchez FM, Vallejo-Cremades MT, de Segura IA, del Campo Ede M. "Direct activation of telomerase by GH via phosphatidylinositol 3'-kinase." *J Endocrinol.* 2005 Jun; 185(3): 421–8.

## 8 | Age-Proof Your Lifestyle

1 Ludlow, Andrew T, and Roth SM. "Physical activity and telomere biology: exploring the link with aging-related disease prevention." *J Aging Res.* 2011 Feb 21; 2011: 790378. doi: 10.4061/2011/790378.

2 Cherkas, Lynn F, Hunkin JL, Kato BS, Richards JB, Gardner JP, Surdulescu GL, Kimura M, Lu X, Spector TD, Aviv A. "The association between physical activity in leisure time and leukocyte telomere length." *Arch Intern Med.* 2008 Jan 28; 168(2): 154–8. doi: 10.1001/archinternmed.2007.39.

3 Casey, A, Constantin-Teodosiu D, Howell S, Hultman E, Greenhaff PL. "Creatine ingestion favorably affects performance and muscle metabolism during maximal exercise in humans." *Am J Physiol.* 1996 Jul; 271 (1 Pt 1): E31–7.

4 Kreider, RB, Ferreira M, Wilson M, Grindstaff P, Plisk S, Reinardy J, Cantler E, Almada AL. "Effects of creatine supplementation on body composition, strength, and sprint performance." *Med Sci Sports Exerc.* 1998 Jan; 30(1): 73–82.

5 Collins, M, Renault V, Grobler LA, St Clair Gibson A, Lambert MI, Wayne Derman E, Butler-Browne GS, Noakes TD, Mouly V. "Athletes with exercise-associated fatigue have abnormally short muscle DNA telomeres." *Med Sci Sports Exerc.* 2003 Sep; 35(9): 1,524–28.

6 Rae, Dale E, Vignaud A, Butler-Browne GS, Thornell LE, Sinclair-Smith C, Derman EW, Lambert MI, Collins M. "Skeletal muscle telomere length in healthy, experienced, endurance runners." *Eur J Appl Physiol.* 2010 May; 109(2): 323–30. doi: 10.1007/s00421-010-1353-6.

7 LaRocca, Thomas J., Seals DR, Pierce GL. "Leukocyte telomere length is preserved with aging in endurance exercise-trained adults and related to maximal aerobic capacity." *Mech Ageing Dev.* 2010 Feb; 131(2): 165–7. doi: 10.1016/j.mad.2009.12.009.

8 Ponsot, Elodie, Lexell J, Kadi F. "Skeletal muscle telomere length is not impaired in healthy physically active old women and men." *Muscle Nerve.* 2008 Apr; 37(4): 467–72. doi: 10.1002/mus.20964.

9 Ji, Li Li. "Modulation of skeletal muscle antioxidant defense by exercise: Role of redox signaling." *Free Radic Biol Med.* 2008 Jan 15; 44(2): 142–52. doi: 10.1016/j.freeradbiomed.2007.02.031.

10 Ji, LiLi, Gomez-Cabrera MC, Vina J. "Exercise and hormesis: activation of cellular antioxidant signaling pathway." *Ann N Y Acad Sci.* 2006 May; 1067: 425–35.

11 Büchner, Nicole, Zschauer TC, Lukosz M, Altschmied J, Haendeler J. "Downregulation of mitochondrial telomerase reverse transcriptase induced by $H_2O_2$ is Src kinase dependent." *Exp Gerontol.* 2010 Aug; 45(7-8): 558–62. doi: 10.1016/j.exger.2010.03.003.

12 Haendeler, Judith, Dröse S, Büchner N, Jakob S, Altschmied J, Goy C, Spyridopoulos I, Zeiher AM, Brandt U, Dimmeler S. "Mitochondrial telomerase reverse transcriptase binds to and protects mitochondrial DNA and function from damage." *Arterioscler Thromb Vasc Biol.* 2009 Jun; 29(6): 929–35. doi: 10.1161/ATVBAHA.109.185546.

13 Kadi, F, Ponsot E, Piehl-Aulin K, Mackey A, Kjaer M, Oskarsson E, Holm L. "The effects of regular strength training on telomere length in human skeletal muscle." *Med Sci Sports Exerc.* 2008 Jan; 40(1): 82–7.

14 Savage, Luiza Ch. "Sleep crisis: The science of slumber." [*Maclean's* website]. June 17, 2013. Available at macleans.ca/2013/06/17/the-sleep-crisis/. Accessed February 19, 2014.

15 Varughese, Jason, Allen, Richard P. "Fatal accidents following changes in daylight savings time: the American experience." *Sleep Med.* 2001 Jan; 2(1): 31–6.

16 Coren, S. "Daylight savings time and traffic accidents." *N Engl J Med.* 1996 Apr 4; 334(14): 924.

17 "Heart attacks rise following daylight saving time." March 7, 2012. [Science Daily website]. Available at sciencedaily.com/releases/2012/03/120307162555.htm. Accessed February 19, 2014.

18 Morin. CM, LeBlanc M, Bélanger L, Ivers H, Mérette C, Savard J. "Prevalence of insomnia and its treatment in Canada." *Can J Psychiatry.* 2011 Sep; 56(9): 540–8.

19 Leproult, R, Copinschi G, Buxton O, Van Cauter E. "Sleep loss results in an elevation of cortisol levels the next evening." *Sleep.* 1997; 20(10): 865–70 (31 ref.).

20 Institute of Medicine (US) Committee on Sleep Medicine and Research; Colten HR, Altevogt BM, editors. Washington (DC): National Academies Press (US); 2006.

21 Savage, Luiza Ch. Sleep crisis: The Science of Slumber. [*Maclean's* website]. June 17, 2013. Available at macleans.ca/2013/06/17/the-sleep-crisis/. Accessed February 19, 2014.

22 Knutson, Kristen L, Spiegel K, Penev P, Van Cauter E. "The metabolic consequences of sleep deprivation." *Sleep Med Rev.* 2007 Jun; 11(3): 163–78.

23 Cappuccio, Francesco P, D'Elia L, Strazzullo P, Miller MA. "Sleep duration and all-cause mortality: a systematic review and meta-analysis of prospective studies." *Sleep.* 2010 May; 33(5): 585–92.

24 Jackowska, Marta, Hamer M, Carvalho LA, Erusalimsky JD, Butcher L, Steptoe A. "Short sleep duration is associated with shorter telomere length in healthy men: findings from the Whitehall II cohort study." *PLoS One.* 2012; 7(10): e47292. doi: 10.1371/journal.pone.0047292.

25 Prather, Aric A, Puterman E, Lin J, O'Donovan A, Krauss J, Tomiyama AJ, Epel ES, Blackburn EH. "Shorter leukocyte telomere length in midlife women with poor sleep quality." *J Aging Res.* 2011; 2011: 721390. doi: 10.4061/2011/721390.

26 Thakkar, Mahesh M, Sharma R, Sahota P. "Alcohol disrupts sleep homeostasis." *Alcohol Online.* 2014; doi: 10.1016/j.alcohol.2014.07.019.

27 Watkins ES. *The Estrogen Elixir: A History of Hormone Replacement Therapy in America.* Baltimore: John Hopkins University Press; 2007.

28 Pierpaoli, Walter, Regelson, William, Colman, Carol. *The Melatonin Miracle: Nature's Age-Reversing, Disease-Fighting, Sex-Enhancing Hormone.* New York City: Simon & Schuster Inc.; 1996.

29 Irwin, M, McClintick J, Costlow C, Fortner M, White J, Gillin JC. "Partial night sleep deprivation reduces natural killer and cellular immune responses in humans." *FASEB J.* 1996 Apr; 10(5): 643–53.

30 Barceló, Antonia, Piérola J, López-Escribano H, de la Peña M, Soriano JB, Alonso-Fernández A, Ladaria A, Agustí A. "Telomere shortening in sleep apnea syndrome." *Respir Med.* 2010 Aug; 104(8): 1,225–29. doi: 10.1016/j.rmed.2010.03.025.

31 Pavanello, Sofia, Hoxha M, Dioni L, Bertazzi PA, Snenghi R, Nalesso A, Ferrara SD, Montisci M, Baccarelli A. "Shortened telomeres in individuals with abuse in alcohol consumption." *Int J Cancer.* 2011 Aug 15; 129(4): 983–92. doi: 10.1002/ijc.25999.

32 Bagnardi, V, Rota M, Botteri E, Tramacere I, Islami F, Fedirko V, Scotti L, Jenab M, Turati F, Pasquali E, Pelucchi C, Bellocco R, Negri E, Corrao G, Rehm J, Boffetta P, La Vecchia C. "Light alcohol drinking and cancer: a meta-analysis." *Ann Oncol.* 2013 Feb; 24(2): 301–8. doi: 10.1093/annonc/mds337.

33 Vachon, CM, Cerhan JR, Vierkant RA, Sellers TA. "Investigation of an interaction of alcohol intake and family history on breast cancer risk in the Minnesota Breast Cancer Family Study." *Cancer.* 2001 Jul 15; 92(2): 240–8.

34 Chen, Wendy Y., Rosner B, Hankinson SE, Colditz GA, Willett WC. "Moderate alcohol consumption during adult life, drinking patterns, and breast cancer risk." *JAMA.* 2011 Nov 2; 306(17): 1,884–90. doi: 10.1001/jama.2011.1590.

35 Valdes AM, Andrew T, Gardner JP, Kimura M, Oelsner E, Cherkas LF, Aviv A, Spector TD. "Obesity, cigarette smoking, and telomere length in women." *Lancet.* 2005 Aug 20-26; 366(9486): 662–4.

36 van der Vaart, H, Postma DS, Timens W, ten Hacken NH. "Acute effects of cigarette smoke on inflammation and oxidative stress: a review." *Thorax.* 2004 Aug; 59(8): 713–21.

## 9 | Mental Health

1 Harley, Calvin, and Epel, Elissa. "Telomeres and Aging: how stress causes aging." Filmed 2011. TedMed Talk, 12:41. Posted January 2012. youtube.com/watch?v=mnMuuRcczf8.

2 Wolkowitz, Owen M, Mellon SH, Epel ES, Lin J, Dhabhar FS, Su Y, Reus VI, Rosser R, Burke HM, Kupferman E, Compagnone M, Nelson JC, Blackburn EH. "Leukocyte telomere length in major depression: correlations with chronicity, inflammation and oxidative stress—preliminary findings." *PLoS One*. 2011 Mar 23; 6(3): e17837. doi: 10.1371/journal.pone.0017837.

3 Puterman, Eli, and Epel E. "An intricate dance: Life experience, multisystem resiliency, and rate of telomere decline throughout the lifespan." *Soc Personal Psychol Compass*. 2012 Nov 1; 6(11): 807–25. Epub 2012 Nov 5.

4 Entringer, Sonja, Epel ES, Kumsta R, Lin J, Hellhammer DH, Blackburn EH, Wüst S, Wadhwa PD. "Stress exposure in intrauterine life is associated with shorter telomere length in young adulthood." *Proc Natl Acad Sci U S A*. 2011 Aug 16; 108(33): E513–8. doi: 10.1073/pnas.1107759108.

5 Surtees, Paul G, Wainwright NW, Pooley KA, Luben RN, Khaw KT, Easton DF, Dunning AM. "Life stress, emotional health, and mean telomere length in the European Prospective Investigation into Cancer (EPIC)-Norfolk population study." *J Gerontol A Biol Sci Med Sci*. 2011 Nov; 66(11): 1,152–62. doi: 10.1093/gerona/glr112.

6 Folkman, Susan. "Personal control and stress and coping processes: A theoretical analysis." *Journal of Personality and Social Psychology*, 1984 Apr; 46(4): 839–852.

7 Wallace, B Allen. *Balancing the Mind: A Tibetan Buddhist Approach to Refining Attention*. Ithaca, New York: Snow Lion Publications; 2005.

8 Baer, RA, Smith GT, Hopkins J, Krietemeyer J, Toney L. "Using self-report assessment methods to explore facets of mindfulness." *Assessment*. 2006 Mar; 13(1): 27–45.

9 Ornish, Dean, Lin J, Daubenmier J, Weidner G, Epel E, Kemp C, Magbanua MJ, Marlin R, Yglecias L, Carroll PR, Blackburn EH. "Increased telomerase activity and comprehensive lifestyle changes: a pilot study." *Lancet Oncol*. 2008 Nov; 9(11): 1,048–57. doi: 10.1016/S1470-2045(08)70234-1.

10 Daubenmier, Jennifer, Lin J, Blackburn E, Hecht FM, Kristeller J, Maninger N, Kuwata M, Bacchetti P, Havel PJ, Epel E. "Changes in stress, eating, and metabolic factors are related to changes in telomerase activity in a randomized mindfulness intervention pilot study." *Psychoneuroendocrinology.* 2012 Jul; 37(7): 917-28. doi: 10.1016/j.psyneuen.2011.10.008.

11 Lavretsky, H, Epel ES, Siddarth P, Nazarian N, Cyr NS, Khalsa DS, Lin J, Blackburn E, Irwin MR. "A pilot study of yogic meditation for family dementia caregivers with depressive symptoms: effects on mental health, cognition, and telomerase activity." *Int J Geriatr Psychiatry.* 2013 Jan; 28(1): 57–65. doi: 10.1002/gps.3790.

12 Jacobs, Tonya L, Epel ES, Lin J, Blackburn EH, Wolkowitz OM, Bridwell DA, Zanesco AP, Aichele SR, Sahdra BK, MacLean KA, King BG, Shaver PR, Rosenberg EL, Ferrer E, Wallace BA, Saron CD. "Intensive meditation training, immune cell telomerase activity, and psychological mediators." *Psychoneuroendocrinology.* 2011 Jun; 36(5): 664–81. doi: 10.1016/j.psyneuen.2010.09.010.

13 Carlson, Linda E, Beattie, TL, Giese-Davis, J, Farris, P, Tamagawa, R, Fick, LJ, Degelman, ES, and Speca, M. "Mindfulness-based cancer recovery and supportive-expressive therapy maintain telomere length relative to controls in distressed breast cancer survivors." *Cancer.* 2014 Nov; 121(3): 2-14. doi: 10.1002/cncr.29063

14 Heppner, Whitney L, Kernis MH. "'Quiet ego' functioning: the complementary roles of mindfulness, authenticity and secure high self-esteem." *Psychol Inq.* 10/2007; 18: 248–51. doi:10.1080/10478400701598330

15 Weinstein, Netta, Brown KW, Ryan RM. "A multi-method examination of the effects of mindfulness on stress attribution, coping, and emotional well-being." *J Res in Pers.* 2009; 43: 374–385. doi: 10.1016/j.jrp.2008.12.008.

16 Epel, Elissa, Daubenmier J, Moskowitz J, Folkman S, Blackburn E. "Can meditation slow rate of cellular aging? Cognitive stress, mindfulness, and telomeres." *Ann N Y Acad Sci.* 2009 August; 1172: 34-53.

17 Dallman, Mary F, Akana SF, Strack AM, Hanson ES, Sebastian RJ. "The neural network that regulates energy balance is responsive to glucocorticoids and insulin and also regulates HPA axis responsivity at a site proximal to CRF neurons." *Ann N Y Acad Sci.* 1995 Dec 29; 771: 730–42.

18 McEwen, Bruce S. "Protective and damaging effects of stress mediators." *N Engl J Med.* 1998 Jan 15; 338(3): 171–9.

19 Adam, Emma K, and Gunnar MR. "Relationship functioning and home and work demands predict individual differences in diurnal cortisol patterns in women." *Psychoneuroendocrinology.* 2001 Feb; 26(2): 189–208.

20 Epel, Elissa, Burke, Heather M., Wokowitz, Owen M. "Psychoneuroendocrinology of Aging: Focus on anabolic and catabolic hormones." In: Aldwin C, Spiro A, Park C, editors. *Handbook of Health Psychology of Aging.* New York: Guildford Press, 2007 (119–41).

21 Hall, Daniel E. "Religious attendance: more cost-effective than lipitor?" *J Am Board Fam Med.* 2006 Mar-Apr; 19(2): 103–9.

22 Schnall, Eliezer, Wassertheil-Smoller S, Swencionis C, Zemon V, Tinker L, O'Sullivan MJ, Van Horn L, Goodwin M. "The relationship between religion and cardiovascular outcomes and all-cause mortality in the women's health initiative observational study." *Psychol Health.* 2010 Feb; 25(2): 249–63. doi: 10.1080/08870440802311322.

23 Bower, Julienne E, Low CA, Moskowitz JT, Sepah S, Epel ES. "Benefit finding and physical health: Positive psychological changes and enhanced allostasis." *Social and Personality Psychology Compass.* 2007; 2: 223–44.

## 10 | Putting Your Peak Health Program Together

1 What Causes Behavior Change? BJ Fogg's Behavior Model. [Stanford University website]. Available at behaviormodel.org. Accessed February 14, 2014.

2 Wood, Wendy, and Neal, David T. "A New Look at Habits and the Habit–Goal Interface." *Psychol Rev.* 2007; 114(4): 843–63.

3 Graybiel, Ann M. "Habits, rituals, and the evaluative brain." *Annu Rev Neurosci.* 2008; 31: 359–87. doi: 10.1146/annurev.neuro.29.051605.112851.

4 Fowler, James H, and Christakis, Nicholas A., *Connected: The Surprising Power of Our Social Networks and How They Shape Our Lives – How Your Friends' Friends' Friends Affect Everything You Feel, Think, and Do.* New York: Little, Brown; 2009.

# Sources

Data for the tables found in this work are taken from the following sources. Entries are listed by page number.

*14* Goklany, Indur M. *The Improving State of the World,* Washington, DC: Cato Institute, 2007.

*15* Historical Statistics of the United States and Statistical Abstract of the United States.

*29* Adapted from "Hayflick, his limit, and cellular ageing" by Shay, J.W., and Wright, W.E. *Nature Revues Molecular Cell Biology,* 2000.

*30* Adapted from "Telomeres and immune competency" by Weng, N.P. *Current Opinion in Immunology,* 2012.

*68* Ronald L. Prior, PhD, chemist and nutritionist, USDA's Arkansas Children's Nutrition Center in Little Rock, Arkansas, in *The Journal of Agricultural and Food Chemistry,* ninth edition, June 2004.

*69* National Institutes of Health, United States Department of Health & Human Services, 2013. ods.od.nih.gov/factsheets/VitaminC-HealthProfessional.

*81* OECD Health Statistics 2014, Organisation for Economic Co-operation and Development, 2014. oecd.org/health/healthdata.

*82* Statistics Canada, CANSIM Table 105-0501; "Prevalence of Overweight, Obesity, and Extreme Obesity Among Adults: United States, Trends 1960–1962 Through 2009–10" by Fryar, Cheryl D., et al. Washington, D.C., National Center for Health Statistics, Division of Health and Nutrition Examination Surveys, 2012.

*83* Statistics Canada; Centers for Disease Control and Prevention.

*93* "Omega-3 Fatty Acid and Health" by Nettleton, Joyce A. New York: Chapman & Hall, 1995; *CRC Handbook of Dietary Fiber in Human Nutrition*, Third Edition by Burkitt, Denis P., and Spiller, Gene A. CRC Press, 2001; Dieticians of Canada, Canadian Nutrient File, 2010.

*98* Adapted from Food and Agriculture Organization of the United Nations, *The State of Food and Agriculture,* 2009.

*109* Health Canada, Canadian Community Health Survey, Cycle 2.2, Nutrition, 2004.

*153* Adapted from "Adult growth hormone deficiency; a higher compliance delivery system" by Braverman, Eric R., et al. *American Academy of Anti-Aging Medicine Tenth International Conference in Las Vegas, Nevada,* 2002.

*160* Adapted from "The association between physical activity in leisure time and leukocyte telomere length" by Cherkas, Lynn F, et al. *Archives of Internal Medicine,* 2008.

*161* Adapted from "The association between physical activity in leisure time and leukocyte telomere length" by Cherkas, Lynn F, et al. *Archives of Internal Medicine,* 2008.

*163* Adapted from "Testosterone, dehydroepiandrosterone, insulin-like growth factor 1, and insulin in sedentary and physically trained aged men" by Tissandier, O., et al. *European Journal of Applied Physiology,* 2001.

*171* Adapted from "Leukocyte telomere length is preserved with aging in endurance exercise-trained adults and related to maximal aerobic capacity" by LaRocca, Thomas J., et al. *Mechanisms of Ageing and Development,* 2010.

*177* Adapted from "Shorter leukocyte telomere length in midlife women with poor sleep quality," by Prather, Aric A., et al. *Journal of Aging Research,* 2011.

*187* Adapted from "Shortened leukocyte telomeres in individuals with abuse in alcohol consumption," by Pavanello, Sofia, et al. *International Journal of Cancer*, 2011.

*188* Adapted from Canadian Centre on Substance Abuse, 2012.

*202* Adapted from "Intensive meditation training, immune cell telomerase activity, and psychological mediators" by Jacobs, T.L., et al. *Psychoneuroendocrinology*, 2010.

*203* Adapted from *The Art of Happiness: A Handbook for Living, tenth Anniversary Edition* by His Holiness the Dalai Lama. New York: Riverhead Books, 2009.

*207* Adapted from "Can meditation slow rate of cellular aging? Cognitive stress, mindfulness, and telomeres" by Epel, Elissa, et al. *Annals of the New York Academy of Sciences*, 2009.

*219* Dr. B.J. Fogg of the Persuasive Technology Lab at Stanford University.

# Index

*Page numbers in italics refer to graphs and illustrations*

RICK O'BRIEN

ELAINE CHIN, MD, MBA, is a North American trailblazer of personalized medicine and is on a mission to educate people on how to stay healthier longer. Using leading edge diagnostic tools to support her goal, she has discovered the power of telomere science to support her claim that everyone can help themselves live longer, perform better—and feel younger too.

She is the founder of Executive Health Centre, a premiere executive health practice in Toronto, Canada, focusing on peak performance through an integrative science-based approach to maintaining optimal health. She works with senior executives and corporations to ensure peak health will lead to peak performance in the workplace. Her clinic's work with telomeres has been highlighted in the *Globe & Mail*, *Maclean's* and *CTV National News*. Her writing has appeared in *Maclean's* and *Canadian Business*.

She is also a consultant for TELUS Corporation, in the role of Chief Wellness Officer. Her primary focus has been to develop and promote innovative health and wellness programs and incentives for the TELUS team and Canadians that encourage lasting healthier lifestyles.

Dr. Chin has addressed physician and health care industry conferences in the United States and Europe on the clinical model for individualized preventive health care. In addition, Dr. Chin is a frequent speaker on executive health as a strategic tool in improving workplace productivity and protecting human capital. She has lectured at the Rotman School of Management on integrating prevention-driven models into corporate human resources practice. Presently, she is an adjunct professor at York University's Faculty of Health in Toronto, Canada.

Dr. Chin received her medical degree from the University of Toronto and her MBA from the same university's Rotman School of Management. She lives in Toronto.

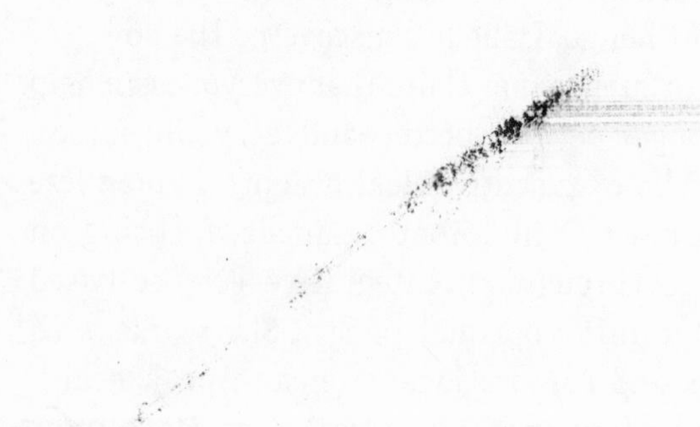